Other Titles by *Langaa* RPCIG

Francis B. Nyamnjoh
Stories from Abakwa
Mind Searching
The Disillusioned African
The Convert
Souls Forgotten
Married But Available

Dibussi Tande
No Turning Back. Poems of Freedom 1990-1993
Scribbles from the Den: Essays on Politics and Collective Memory in Cameroon

Kangsen Feka Wakai
Fragmented Melodies

Ntemfac Ofege
Namondo. Child of the Water Spirits
Hot Water for the Famous Seven

Emmanuel Fru Doh
Not Yet Damascus
The Fire Within
Africa's Political Wastelands: The Bastardization of Cameroon
Oriki'badan
Wading the Tide
Stereotyping Africa: Surprising Answers to Surprising Questions

Thomas Jing
Tale of an African Woman

Peter Wuteh Vakunta
Grassfields Stories from Cameroon
Green Rape: Poetry for the Environment
Majunga Tok: Poems in Pidgin English
Cry, My Beloved Africa
No Love Lost
Straddling The Mungo: A Book of Poems in English & French

Ba'bila Mutia
Coils of Mortal Flesh

Kehbuma Langmia
Titabet and the Takumbeng
An Evil Meal of Evil

Victor Elame Musinga
The Barn
The Tragedy of Mr. No Balance

Ngessimo Mathe Mutaka
Building Capacity: Using TEFL and African Languages as Development-oriented Literacy Tools

Milton Krieger
Cameroon's Social Democratic Front: Its History and Prospects as an Opposition Political Party, 1990-2011

Sammy Oke Akombi
The Raped Amulet
The Woman Who Ate Python
Beware the Drives: Book of Verse
The Wages of Corruption

Susan Nkwentie Nde
Precipice
Second Engagement

Francis B. Nyamnjoh &
Richard Fonteh Akum
The Cameroon GCE Crisis: A Test of Anglophone Solidarity

Joyce Ashuntantang & Dibussi Tande
Their Champagne Party Will End! Poems in Honor of Bate Besong

Emmanuel Achu
Disturbing the Peace

Rosemary Ekosso
The House of Falling Women

Peterkins Manyong
God the Politician

George Ngwane
The Power in the Writer: Collected Essays on Culture, Democracy & Development in Africa

John Percival
The 1961 Cameroon Plebiscite: Choice or Betrayal

Albert Azeyeh
Réussite scolaire, faillite sociale : généalogie mentale de la crise de l'Afrique noire francophone

Aloysius Ajab Amin & Jean-Luc Dubois
Croissance et développement au Cameroun : d'une croissance équilibrée à un développement équitable

Carlson Anyangwe
Imperialistic Politics in Cameroun: Resistance & the Inception of the Restoration of the Statehood of Southern Cameroons
Betrayal of Too Trusting a People: The UN, the UK and the Trust Territory of the Southen Cameroons

Bill F. Ndi
K'Cracy, Trees in the Storm and Other Poems
Map: Musings On Ars Poetica
Thomas Lurting: The Fighting Sailor Turn'd Peaceable /Le marin combattant devenu paisible

Kathryn Toure, Therese Mungah Shalo Tchombe & Thierry Karsenti
ICT and Changing Mindsets in Education

Charles Alobwed'Epie
The Day God Blink...
The Bad C...

... bicycle, taxi and
... Mzuzu, Malawi

Justice Nyo' Wakai
Under the Broken Scale of Justice: The Law and My Times

John Eyong Mengot
A Pact of Ages

Ignasio Malizani Jimu
Urban Appropriation and Transformation: Bicycle Taxi and Handcart Operators

Joyce B. Ashuntantang
Landscaping and Coloniality: The Dissemination of Cameroon Anglophone Literature

Jude Fokwang
Mediating Legitimacy: Chieftaincy and Democratisation in Two African Chiefdoms

Michael A. Yanou
Dispossession and Access to Land in South Africa: an African Perspevctive

Tikum Mbah Azonga
Cup Man and Other Stories
The Wooden Bicycle and Other Stories

John Nkemngong Nkengasong
Letters to Marions (And the Coming Generations)

Amady Aly Dieng
Les étudiants africains et la littérature négro-africaine d'expression française

Tah Asongwed
Born to Rule: Autobiography of a life President

Frida Menkan Mbunda
Shadows From The Abyss

Bongasu Tanla Kishani
A Basket of Kola Nuts

Fo Angwafo III S.A.N of Mankon
Royalty and Politics: The Story of My Life

Basil Diki
The Lord of Anomy

Churchill Ewumbue-Monono
Youth and Nation-Building in Cameroon: A Study of National Youth Day Messages and Leadership Discourse (1949-2009)

Emmanuel N. Chia, Joseph C. Suh & Alexandre Ndeffo Tene
Perspectives on Translation and Interpretation in Cameroon

Linus T. Asong
The Crown of Thorns
No Way to Die
A Legend of the Dead: Sequel of *The Crown of Thorns*
The Akroma File
Salvation Colony: Sequel to *No Way to Die*

Vivian Sihshu Yenika
Imitation Whiteman

Beatrice Fri Bime
Someplace, Somewhere
Mystique: A Collection of Lake Myths

Shadrach A. Ambanasom
Son of the Native Soil
The Cameroonian Novel of English Expression: An Introduction

Tangie Nsoh Fonchingong and Gemandze John Bobuin
Cameroon: The Stakes and Challenges of Governance and Development

Tatah Mentan
Democratizing or Reconfiguring Predatory Autocracy? Myths and Realities in Africa Today

Roselyne M. Jua & Bate Besong
To the Budding Creative Writer: A Handbook

Albert Mukong
Prisonner without a Crime: Disciplining Dissent in Ahidjo's Cameroon

Mbuh Tennu Mbuh
In the Shadow of my Country

Bernard Nsokika Fonlon
Genuine Intellectuals: Academic and Social Responsibilities of Universities in Africa

Lilian Lem Atanga
Gender, Discourse and Power in the Cameroonian Parliament

Cornelius Mbifung Lambi & Emmanuel Neba Ndenecho
Ecology and Natural Resource Development
in the Western Highlands of Cameroon: Issues in Natural Resource Managment

Ecology and Natural Resource Development in the Western Highlands of Cameroon

Issues in Natural Resource Management

Cornelius Mbifung Lambi
&
Emmanuel Neba Ndenecho

Langaa Research & Publishing CIG
Mankon, Bamenda

Publisher:
Langaa RPCIG
Langaa Research & Publishing Common Initiative Group
P.O. Box 902 Mankon
Bamenda
North West Region
Cameroon
Langaagrp@gmail.com
www.langaa-rpcig.net

Distributed outside N. America by African Books Collective
orders@africanbookscollective.com
www.africanbookscollective.com

Distributed in N. America by Michigan State University Press
msupress@msu.edu
www.msupress.msu.edu

ISBN: 9956-615-48-X

DISCLAIMER

All views expressed in this publication are those of the author and do not necessarily reflect the views of Langaa RPCIG.

Content

The Editors

Professor Cornelius Mbifung Lambi holds a Bachelor of Science Degree (General) in Geology and Geography, and a Master of Science Degree in Geography with specialization in Geomorphology from Fourah Bay College of the University of Sierra Leone, Freetown, and a Doctor of Philosophy in Physical Geography from the University of Salford, Greater Manchester University, Salford, Lancashire, England. Professor Lambi is the Former Vice-Chancellor of the University of Buea, Cameroon.

Professor Lambi taught for 17 years in Cameroon's mother University in Yaounde prior to his transfer to the University of Buea in 1993, where he served as the pioneer Head of Geography Department from 1993-2000. He was appointed Dean of the Faculty of Social and Management Sciences in July 1997, a position he held until September 2005, when he was appointed Vice-Chancellor of the University of Buea. He is a Professor of Geography.

He has been a Visiting International Scholar in a number of United States Colleges and Universities. For the 1984–85 academic year, he was Visiting Scholar in the College of Earth and Mineral Sciences at Pennsylvania State University, State College, and during the spring of 1996, he was also a Visiting International Scholar at Dickinson College, Carlisle, Pennsylvania, USA. In addition, in 1994, under the Overseas Development Agencies, he was on attachment to the Environmental Unit of the Faculty of Education of the Jordanhill Campus of the University of Strathclyde, Glasgow, Scotland.

He has published extensively in learned academic journals. He is member of the Editorial Board for a number of scholarly journals, and Chief Editor of the Journal of Applied Social Sciences. He is the Editor of *Readings In Geography* and *Environmental Issues: Problems and* Prospects.

Professor Lambi is a Fellow of the Royal Geographical Society and Fellow of Geological Society of London. He is also a member of the Geological Society of France. He is a consultant on environmental issues.

Dr. Ndenecho Emmanuel Neba who is a Senior Lecturer in Geography at the University of Yaounde 1 (E.N.S. Annex Bambili) holds a Licence and Maitrise in Geography from the University of Yaoundé, and an M.Sc in Land Resource Management from Cranfield University (England), and a specialist Diploma in Rural Physical Planning for Soils and Water Conservation from the Ruppin Institute (Israel). He obtained a doctorate in Geography from the University of Buea.

His previous teaching was in the Cameroon Ministry of Agriculture where he was trainer/director of the Regional College of Agriculture-Bambili, and trainer/director of the National Cooperative College in Bamenda. His special interests and experience include research and development associated with natural resource management, sustainable livelihoods and landscape ecological problems. Originally trained as a geographer, he specialised in Natural Resource Management and apart from university teaching; he is also a consultant on natural resource management and rural livelihood issues. Dr. Ndenecho is also the author of the following books:

- Biological Exploitation in Cameroon: From Crisis to Sustainable Management.
- Sustainable Mountain Environments and Rural Livelihoods in Bamenda highlands, Cameroon.
- The Population – Resource and Conflict Trinity: Analysis of North West Cameroon (co-author).
- Upstream Water Resource Management Strategy and Stakeholder Participation: Lessons in North West Cameroon.
- Landslides and Torrent-Channel Problems of Mountain Slopes: Processes and Management Options for Bamenda Highlands.

List of Acronyms and Abbreviations

C.D.C.	Cameroon Development Corporation
CEMAC	Communauté Economique et Monétaire d'Afrique Centrale (Central African Economic and Monitory Community
GPS	Global Positioning System
ICBP	International Council for Bird Preservation
ICUN	International Union for the Conservation of Nature and Natural Resources
IMFP	Ijim Mountain Forest Project
ITCZ	Inter-tropical Convergence Zone
KMFP	Kilum Mountain Forest Project
MIDENO	Mission de Développement de la Province de Nord Ouest (North West Development Authority)
MINPAT	Ministry of Planning and Regional Development
NGOs	Non-Governmental Organisations
NTFPs	Non-Timber Forest Products
SATA	Swiss Association for Technical Assistance
SEDA	French Acronym for Research Organisation in Cameroon
SNEC	National Water Corporation
SWOT	Strengths, Weaknesses, Opportunities and Threats
SWP	Scan Water Projects
UDEAC	Union de Développement Economique d'Afrique Centrale (Central African Economic Development Union)
UNDP	United Nations Development Programme
UNVDA	Upper Nun Valley Development Authority
WWF	World Wildlife Fund

Chapter One

A Window into the Land Degradation Problem in the North West Province of Cameroon: A Revisit

Summary

Land degradation is an encroaching threat in many parts of the Bamenda mountainous terrain. In an attempt to understand the nature of this phenomenon, this paper examines the different processes and factors which provoke the degradation. These factors relate to the physical, geomorphological and anthropic activities of humans. Although soil erosion is a natural phenomenon, humans, through their reckless use of the land have significantly increased the rate of land degradation. The need for humankind to survive in a world of ever-dwindling resources has been the origin of the conflict between humans and the environment. For some parts of the earth, this anthrogenic degradation has been estimated to be at least two and a half times greater than the natural process of soil regeneration. Mankind has a growing interest in mountain landscapes as some of them appear to be the last vestiges of nature. So humankind in these highlands needs to devise the blueprint for their sustainable development through various appropriate conservation methods.

Introduction

The terms degradation and denudation in geomorphological literature have usually been used synonymously. According to the American Geological Institute (1962), degradation is defined as "the general lowering of the land by erosive processes, especially by the removal of materials through erosion and transportation by flowing water". However, Monkhouse and Small (1978) refer to denudation as the process by which "the earth's surface undergoes destruction, wasting and loss through weathering, mass movement, erosion and transportation".

This study has been undertaken in order to bring into focus the impact of a silent environmental problem that is threatening some of our heavily populated and highly humanised highland regions of Cameroon (Fig 1). In this way, it is hoped that continuously raising awareness amongst mountain stakeholders that the careful and efficient utilisation of their highland resources would improve their sustainable exploitation. Furthermore, in recognition of this environmental threat to the home of an indigenous people, the government, the community as well as non-governmental organisations (NGOs) might design and support some effective methods to mitigate land degradation. Since in the natural scheme of things humans should remain conservators of our natural resources rather than despoilers of our regional or planetary endowments, this paper advocates the need for the proper management of our fragile mountainous landscape.

Fig. 1: Location of the Bamenda Highlands of Cameroon

Indeed, the future of these high-energy environments and the well-being of the indigenous people are threatened by the increasing demands of the growing population as well as the poor management of the existing natural resources. This study also aims at advocating sustainable soil, water and vegetation conservation and management since the survival of future generations in this dynamic highland can only be ensured if we halt the widespread deterioration of the overburdened soil resources. Although many parts of this highland have been destroyed, the further spread of environmental degradation must be checked if the highland home for mankind today must remain a habitably productive place for future generations in the years that lie ahead. After all, mountains have been described as the water towers of the world, vital to all life on earth and to the well-being of people everywhere and it has been said that what happens on the highlands also affects life in the lowlands (Awake! March 22, 2005).

The Problem

As land degradation attains a conspicuously high magnitude in the North West Province of Cameroon, this research is intended to address three major questions which can provide a window into the problem of environmental deterioration. The questions which relate to land use are examined by the help of the following assumptions:

i) There is a direct relationship between rising population density and accelerated soil erosion.
ii) The pursuit of economic gains through the clearing of land for intensive agriculture and the abusive use of rangeland are some of the factors which stand at the forefront of land degradation.
iii) Apart from anthropogenic influences, the physical milieu per se provides a proper setting for a high magnitude of land degradation.

The Environmental Traumas

The tropical highlands of the North West Province, generally referred to as the Bamenda Highlands, with high-altitude monsoonal climate, consist of unbroken mountain ranges characterised by steep slopes and deep valleys. Parts of these chains have abrupt escarpments, towering volcanic peaks, deep valleys and broad undulating alluvial plains. Mount Oku, at an elevation of 3008m above sea level, is the highest in these highlands. Against all expectations, these are the same hostile and fragile landscapes which have been intensely humanised and converted to a cultural landscape. Land in the North West Province of Cameroon is constantly under greater pressure from deforestation and a rapidly growing population. By opening up new farmlands and grazing lands, the natural vegetation is altered, thereby exposing the soils to the ugly sculpting hands of erosion and degradation.

With increasing human interferences on the hilly land surfaces in the Bamenda region through intensive land use, several hectares of agriculturally fertile soils are continuously being lost over the years through soil erosion. It is therefore paradoxical that in the wake of a rapidly increasing population in this highland area, there is falling soil fertility. When the topographical setting (relief) and the population density maps of this region are compared (Fig 2A & B), it is evident that most of the high densities predominantly coincide with the high-altitude zones. Consequently, the livelihoods of humans in these high-altitude areas are strongly tied to the environment itself. So as fragile mountainous ecosystems, these highlands need to be protected as the home of humans.

With a total land surface of 17300sq km, this province had a population of about 1.350.000 people (1987 census) giving an average density of about 80 persons per sq km. The demographic projections for the North West Region of Cameroon for 2010 put the population at 2.090.300 (Ministry of Planning and Regional Development (MINPAT) United Nations Development Programme (UNDP, 1999). This gives an average population density of more that 120 person/sq km. Such high densities are much akin to those of the highlands of Rwanda and Burundi, which register some of the highest population densities in Sub-Saharan Africa. However, some high-density areas around Bamenda, the principal city of the province, have over 200 persons per sq km. This state of affairs is a cause for concern, and it is a signal to all mankind that something must be done urgently to avoid future food problems particularly in this highland region, where the application of modern agricultural inputs such as fertilisers is a remote idea in some areas.

Moreover, at peasant subsistence level, the cost of fertilisers is simply prohibitive. Not blessed with other natural resources like mineral wealth from the bosom of its underlying substratum as is the case in some parts of the world, humans in this highland depends on agriculture for his livelihood; the intensive land use on many hilly slopes has led to man-made land degradation, leading to a loss of agricultural productivity. Some disquieting examples of highly eroded and unproductive slopes can be seen in parts of the Kom plateau, the Oku-Kilum mountain zone and much of the Banso and Nkambe Plateaux. Given this severe state of soil erosion, today, we perhaps share the worries of Edouard Saouma (1933), the Director of the Food and Agricultural Organisation of the United Nations (FAO) who entertains doubts concerning soil agricultural productivity. After a careful "analysis of the man-made land degradation" he asked a fundamental question about the earth's capacity to provide agriculturally fertile lands for growing sufficient food to take care of the frightful galloping population growth by the turn of the century. His fears about humankind's inability to procure food for survival in the next century have been expressed as follows:

> *"Are we going to have enough land to feed the extra 2.6 billion people who will be on this planet by the year 2025?*

In the Bamenda highland region where steep slopes and an aggressive climate (torrential rainfall, high rainfall intensity and high temperatures) predominate, an urgent solution to the problem of erosion is more vital than ever before lest the swarming teams of humankind and cattle be exposed to the devastating consequences of food shortages in the not too distant future.

The loss of fertile topsoil which has been described as "humanity's most precious physical resource" through mismanagement and erosion has been marked in many parts of this province by the use of improved crop varieties as well as the application of chemical fertilisers. Because of the increasing demographic pressure, "marginal lands and steep hillsides (Fig.3) are being ploughed up, overgrazed or stripped of timber and vegetation. These fragile lands quickly erode and lose fertility". Where erosion is advanced, the soils on some of these marginal lands become converted from a renewable to a non-renewable resource.

Fig. 2A: Relief of the Bamenda Highlands

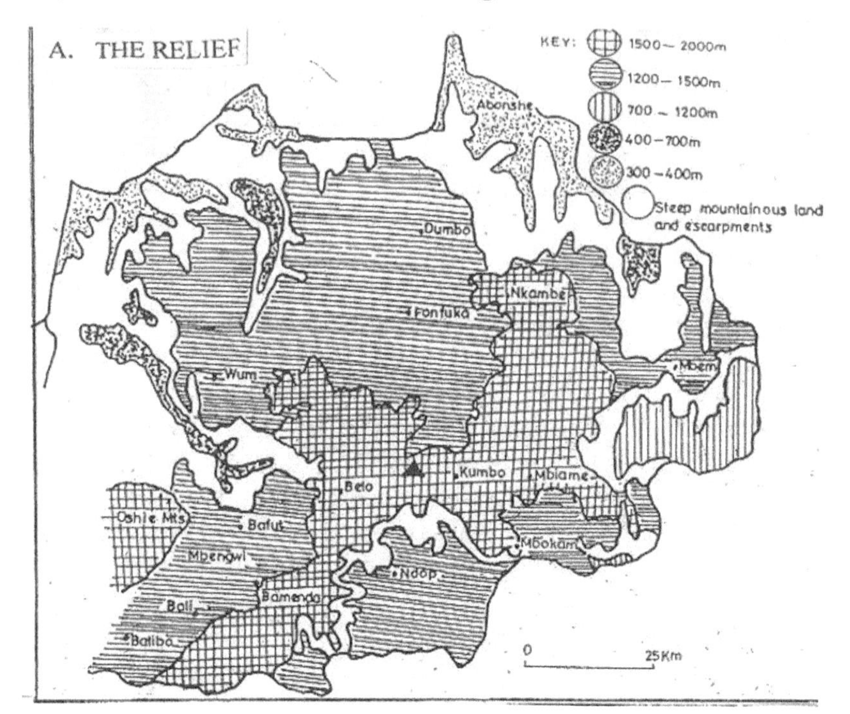

In the Republic of Cameroon, the Bamenda Highlands and the Bamboutous Mountains (Fig.1) remain some of the regions with the greatest land degradation problems. This is so for three basic reasons, namely:

i) the aggressively dynamic climate of the highland
ii) the hilly nature or topographic layout of the land, and
iii) human interference through overgrazing, destructive agricultural practices and the impact of deforestation.

Even on a global scale, the falling food output and food crisis are blamed on environmental factors, foremost among which are deteriorating soils, water and climatic conditions. Holding centre stage amogst the environmental factors responsible for the falling food output is the deteriorating soils. Perhaps it is time for people in this highland region to examine critically the suggestion of Lester

Fig. 2B: Population Density of the Bamenda Highlands

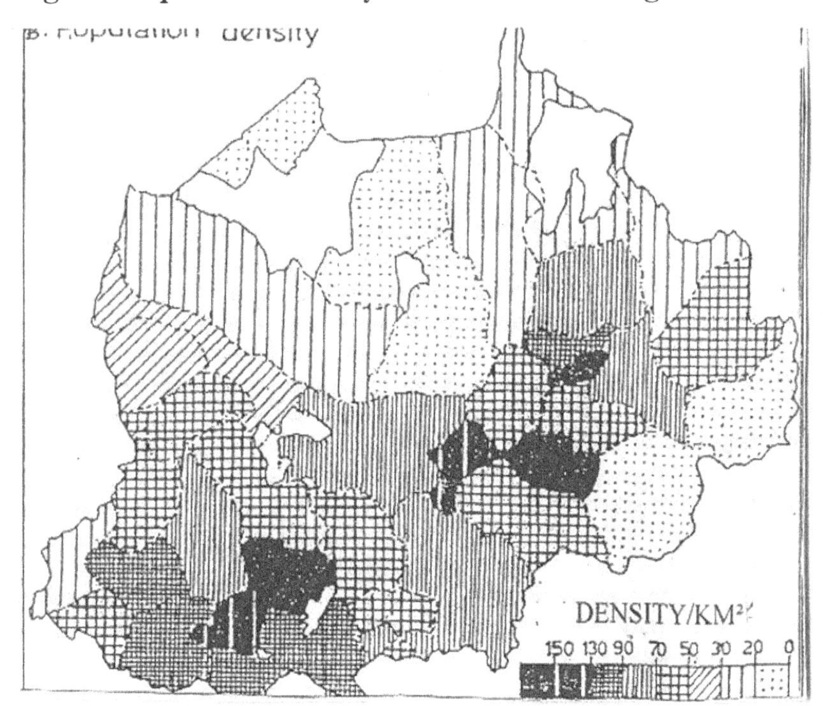

Brown (Worldwatch Institute, Oct. 1988) who claims that we are feeding ourselves at the expense of our children "by ploughing up too much land that is vulnerable to erosion."

Agriculture is the most dynamic of all the components of the economy of the North West Province. While most areas in this highland region enjoy food abundance, the inhabitants of a handful of infertile, rugged mountainous pockets still nurse the fear of periodic or short-term annual food shortages. So one of humanity's primary concerns today relates to the region's future ability to feed itself if the population continues to grow at the present rate of more than 2% per annum. With double cropping every year, humanity's intensified land exploitation leaves no room for soil regeneration in parts of the Kom Plateau which is one of the high-density mountainous regions. The population problem in relation to land degradation is infinitely more complicated than the region's sheer ability to produce enough food supply.

Fig. 3: Cultivated Marginal Lands at Memfu (Bui Division)

Historical Review

With a smaller human population, and before the arrival of the Fulani graziers in these highlands some 90 to 100 years ago, only relatively small portions of the grasslands and available arable land were utilised. Since there was limited pressure on land, the soil could recover its fertility after the periodic land use and long fallow systems of yesteryear.

The present high population concentrations, however, deplete the soil faster than the earth can possibly heal over or recover its fertility, especially on the steep and rugged slopes for which much of this region is reputed. Since the coming of the Fulani graziers, larger and larger cattle populations have become the order of the day, and the upland areas of this region are now used for extensive wet-season grazing. Cattle transhumance from the dry, burnt-up hills from December to March takes place towards the plains and the watered intermontane valleys.

Cattle grazing in this region is one of the major factors responsible for pasture degradation. While arable land accounts for about 40 – 45% of the total land surface, grazing lands are estimated at 55–60%. For long historic periods, the North West grasslands and mountainous backbone have provided pastures or the commons for all herdsmen. Their increasing cattle numbers have continued to find abundant room for grazing on these commons, which are public property. This overgrazing of the commons with its consequent economic ruin has not been a healthy exercise from the perspective of environmental conservation. Perhaps an excellent example of the negative environmental legacy from cattle trampling and land degradation had been provided by the numerous terracettes near the Jakiri Cattle Market (Bui Division) and other areas.

Allo (1996) gave the livestock stocking rate for the North West Province as 2-4 ha/head/year. However, using 1998 livestock figures and with the recognition that the total grazing land has remained the same at 10599sq km, the projected stocking rate would be estimated at 2.21/ha/head/year. Considering cattle, sheep and goats together, the stocking rate drops to nearly 2.08 ha/head/year. With the increasing demographic pressure for this province, the stocking rate for the dwindling grazing lands is thus probably much higher. This livestock pressure provides the *raison d'etre* for the observed overgrazing and pasture degradation. The outcome has been land degradation which is a 'tragedy of the commons' (Hardin, 1968). The increase in human population has also engendered cultivation of a larger area of the tropical highland savanna.

The low productivity of the rugged hills and mountain chains makes most of the Kom and some parts of the Oku Highlands and the Bui plateau better suited to cattle grazing. This explains why this mountainous backbone is part of the domain of cattle rearing in the North West Province of Cameroon (Fig.4).

During the period of low population, agriculture was of the shifting type. The mountain inhabitants were mainly farmers who cleared and cultivated portions of the land. This farming process continued until the soil lost its fertility. Such depleted soils were then abandoned for new farm lands which held promise for better yields. The vegetationally naked and abandoned lands in some cases were susceptible to destruction by erosion in this climatically aggressive mountain region.

Fig. 4: Mountainous Chain and Grazing Lands Around Takui (Bui Division)

With the recent population growth, however, the shifting subsistence agriculture has changed into the sedentary and partly commercial production of Arabica coffee. With the former subsistence level, farmers produced the quantity of food needed for the family till the next harvest. However, road modernisation and rapid population growth have upset the former highland subsistence economy, and part of the farm production has become export oriented.

The anthropic modifications of this highland region are many. Thus, man has transformed the highland zone into cultural landscape. The evidence in this respect includes cleared forest lands, contoured and terraced mountainsides within the steep mountainous areas, the colonisation of swampy areas for rice cultivation, and the ploughed-up fields for agriculture.

The vestiges of a once-extensive forested and woodland savanna are many. Enormous quantities of huge timber logs are shipped out of Magba, an adjacent forested area in the Western Province.

However, all that is left today are some concentrated pockets of mountain forests such as the Kilum Mountains Forest Reserve of the Oku, the Bali-Ngemba Forest Reserve, the Kimbi River Fauna Reserve and the North West Forest Reserve along the Katsina Ala Valley extending into Nigeria.

The North West Province state forest (Forest Reserve) covers barely 7% of the total surface area of the Province. In this region, there is a growing demand for timber, fuel wood, agricultural and rangeland and also land for the expanding settlements. These demands have opened up the floodgates to the rapid decline of forest cover in the highland zone. Consequently, forest pockets are progressively retreating (Fig.5) in the face of mounting demographic pressure, increased agricultural activity, the spread of cattle ranching and the excessive exploitation of fuel wood for domestic purposes. This process does not only expose the surface to erosion but also reduces the region's biodiversity. Perhaps the impressive grassland scenery of the region is an eloquent testimony of the trauma and biological holocaust which it has undergone.

Fig. 5: The Retreat Of The Oku Montane Forest from 1963 to 1986

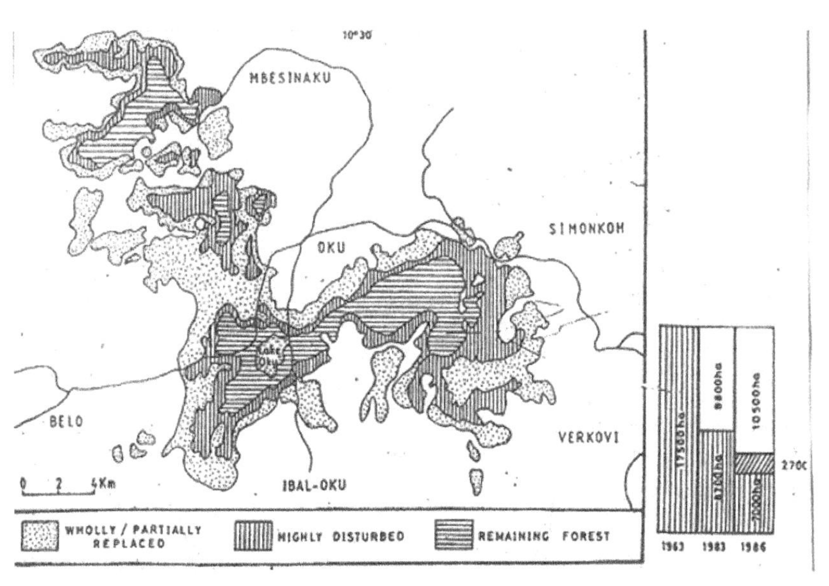

Furthermore, the rape of the overlying vegetal cover exposes the surface to such environmental problems as soils depletion and soil erosion, particularly gullying, given the fact that the North West Province in general is a high-energy environment. The importance of vegetation cover has been stressed by M.O. Hyde (1961), who said that forests hold water and keep the soil in place. "A forest floor is spongy because plants have been growing in it for many years. Without roots in the ground, as much as one inch of soil may run off in a single rain." Yet it takes nature as much as 300 to 1000 years to restore an inch of topsoil. As the connection between forests, soil and water supply has been disrupted in many parts of this mountainous region, bare hills and severely eroded areas stretch over several sectors of the hilly chains. Moreover, the montane forests of the North West Province remain important watersheds from which streams and rivers take their rise.

Certain overexploited areas are scarred by broad and deep erosion gullies; such areas represent zones with the highest concentration of potential erosion, some of which give the crude type of pseudo 'badland topography' such as Ngwa-Ajung around Mbizenaku and Kom Plateau. The former thin layers of humic soils have been washed away by the wet climate since much of the mountainous highland region has higher rainfall intensity. Torrential rainfall lasts from April to September.

Some of the main reasons for soil degradation stem from the widespread deforestation that had been practised, partly to provide wood for domestic purposes such as cooking and heating homes in this altitudinally cold environment. Charcoal burners have also had their part to play. It seems a living anachronism that forest inhabitants (villages were implanted within forested parts of the highlands) moved out for charcoal from the near tree- barren hills of the savanna grasslands, where certain rare species fell victim to the axes of the charcoal burners. Perhaps it would be necessary for humans in these highland areas to cut down on charcoal consumption before there is nothing left for humanity to protect. The pains of rehabilitation through reafforestation are enormous and time-consuming, particularly at this time of galloping population growth.

Since trees and forests, however, go to serve mankind, living things and the environment, they should be rationally exploited so that the existing reserves can sustain future generations. Perhaps the popular slogan "when you cut one tree, plant two" could provide a partial solution to this problem of widespread deforestation in the region, and the consequent loss of biodiversity.

The Fulani and Aku nomads who have roamed these highlands for 9-10 decades also met their firewood needs (energy requirements) by destroying part of the existing vegetation. These cumulative processes laid bare the soils to the capricious effects of an aggressive climate where the sun, rain and mountain wind blew hot and cold. The search for firewood in their open-hearth cooking made the nomads some of the environmental despoilers of this century. After all, these classes of people have no other ways of meeting their energy requirements.

The land use practices in the North West Province are not encouraging as human and livestock numbers are growing rapidly. So in the face of the growing population, the actual rangeland is decreasing. This problem is even further aggravated by the fact that some ecologically fragile zones such as the Kilum Mountain Forest have been demarcated as forest reserves. Land deterioration as a result of overcultivation, deforestation and overgrazing is now widespread in this area.

Given this bleak prospect, much of this settled highland region may, by the turn of the century, become a man-made wasteland with thin, unproductive, barren landscapes. Because the magnitude of soil erosion in the North West Province of Cameroon has given birth to most of the barren, unproductive hills and plateaux, serious environmentalists can aptly look at these highlands as arid lands in transition.

Prospects

What future does this densely populated and humanised mountainous region have for its growing population? No doubt we do at this time require a blueprint for conservation and sustainability because we depend on the earth's resources to meet our vital needs. With the high population densities, the resource base of this mountainous environment is gradually pressed to the limits of its capacity (Fig. 6).

Fig. 6.: Mountainous Slopes in Bui Division Cultivated to the Limits of their Capacity - a Prelude to Degradation

In view of the widespread degradation of forest resources, the Kilum Mountain Forest Project (KMFP) and the Ijim Mountain Forest Project (IMFP) were set up in 1987 in order to conserve the last remnants of this unique and endangered ecosystem. It was estimated that the natural forest for this mountainous chain amounted to some 16,000 hectares of land. Forest conservation has thus protected the forest against destructive farming, burning and grazing. Based on the new philosophy that the forest is meant for man and should be well managed so that the same mountain environments hold promise for future generations, it can be said that man largely depends on the forest for a wide variety of his basic needs. Also as a major geo-hydrological centre, these mountains supply the ground water resources of this region.

In an elaborate study on "Mountain Geography and Resource Conservation", Ndenecho (2006) showed that mountain regions are indeed changing worlds and with particular reference to these

Bamenda Highlands of Cameroon, he focuses on the problems, policies and strategies for sustainable mountain development. The significance of sustainable development has been emphasized by Kofi Annan, former United Nations Secretary General, who said "everyone has a stake in ensuring that the world's mountain regions continue to provide their riches for many generations to come" (Awake!, March 22, 2005).

Certainly, the rapidly rising population is a prelude to further deterioration of the environment. And it seems that a remarkable alteration in the farming methods and new livestock ranching practices in this highland region can hold the promise for halting land degradation.

References

Armstrong, H.W. (Editor) (1985): *The Silent Disaster.*

Brown, L. (1988): *Worldwatch Paper*, No. 24, Worldwatch Institute.

Hardin, G. (1968): *The Tragedy of the Commons*, Science, 162, p 1243.

Hyde, M.O. (1961): *This Crowded Planet*, The New American Library of World Literature, Inc., New York pp 45.

Lambi, C.M. (2001): Environmental Issues: Problems And Prospects. Unique Printers, Bamenda, pp 45-66.

Lambi, C.M. (1999): Land Degradation in the North West Province of Cameroon. In Culture and Environment: A Reader in Environmental Education (Editors) Jim Dunlop & Roy Williams, University of Strathclyde, Glasgow, pp 174-183.

Mountains, Vital For Life on Earth. *AWAKE!*, March 22, 2005.

Ndenecho, E.N. (2006): Mountain Geography and Resources Conservation: Sustaining Mountain Environments and Rural Livelihoods in the Bamenda Highlands, Cameroon. Published in Cameroon, Unique Printers, Bamenda, Cameroon.

The Plain Truth. A Magazine of Understanding, Vol. 50, No. 3, April 1985, p 41.

Vollers, M. (1993): *Topic*, Issue No. 182, p 16

Chapter Two

Environmental Degradation and Problems of Land Resource Management in the Bamenda Highlands, Cameroon

Summary

Mountain environments attract high concentrations of human populations. People also live downstream from mountains and depend on their water, hydropower, grassland, timber and other resources. After several millennia of intensive human transformation of the surrounding lowlands, mountains are one of the last opportunities for conserving natural resources. Unfortunately, many mountain ecosystems face intensive resource extraction. The paper uses a combination of field observation and secondary data to identify the root causes and effects of land degradation. It identifies poor land resource management as the core problem. The main causes include: gender and land ownership problems, population pressure, low prices of staple foods, institutional weaknesses, poor farming and grazing techniques, limited access to credit and training and the effect of bush fires. It also identifies deforestation, the loss of biodiversity, water contamination, low crop yields, water shortages, soil erosion, land disputes, and farmer-grazier conflicts as the effects of land degradation. Finally, the paper elaborates land degradation and management models and identifies the scope for sustainable land resource management in poverty-stricken mountain regions of developing countries. It concludes that poverty and the need to sustain livelihoods are the cause of land degradation.

Introduction

Mountains span 20% of the landscape of the earth and are home to 10% of humanity. An additional 2 billion people live downstream from them; and depend on their water, hydropower, grassland, timber, and mineral resources. Furthermore, 7 of the world's 14 tropical "hotspots" of endemic plants threatened by imminent destruction have at least 50% their area in tropical mountains (Denniston, 1995). The enormous layers of complexity of mountain landscapes their climates, vegetation, and wildlife, have spawned great cultural diversity. For example, several million tribal farmers and pastoralists reside in the mountains of Afghanistan, China, Iran, Nepal, Pakistan, and Central Asian nations of the former Soviet Union.

Around the world, mountain people risk increasing assimilation, debilitating poverty, and political disempowerment. After millennia of intensive human transformation of the surrounding lowlands and flatlands, mountains have become vertical islands of cultural and biological diversity surrounded by seas of biological impoverishment and cultural homogeneity. This enormous diversity, according to Denniston (1995), makes mountains one of the last major opportunities for conserving natural and human variety. Mountain poverty is not a myth; in most parts of the world, mountainous areas are the poorest, most fragile and most vulnerable parts of the country (Khan, 2004). The income level of these areas is often well below the national average. The economy mainly depends on subsistence agriculture and livestock rearing. In such areas, cultivable land is very limited and the majority of the land is barren mountains and steep slopes. There are often very limited on-farm employment opportunities and hence seasonal, cyclical and permanent out-migration of the people to the plain areas of the country in search of jobs is a common phenomenon.

The study seeks to identify the root causes of land degradation and effects on livelihoods in order to establish land degradation and land management models for poor rural communities of sub-Saharan Africa.

The Study Area

The climate is highly varied and is influenced by topography which ranges from an altitude of 300m to 3011m above seal level. It has been described by Moby (1979) as a tropical montane climate characterized by 1500 to 3000mm of rainfall per year, 0 to 3 dry months; a mean annual temperature of 21°c and a mean annual temperature range of 2.2°c. Moist montane forest is the climax vegetation community of the wetter mountains. Lowland evergreen forest is found at elevations below 300m above sea level. These climax floristic communities have been anthropogenically degraded and what exists today is a complex mosaic of montane woodlands, tree and shrub savanna, grass savanna, farms and fallow fields derived from tropical montane forests (Nkwi and Warnier, 1982; Tamura, 1986). In these diverse ecological circumstances, tree and shrub germplasm is extremely varied and reflects to a large extent the differences in ecological factors such as climate, altitude, land use management and edaphic conditions.

The soils are ultosols; poor in major nutrients, acidic and with high phosphorus requirements (Yamoah *et al.* 1994). Furthermore, some food crop fields are found on steep slopes where erosional losses are phenomenal as is the decline in soil fertility. These soils are derived from basalts, trachytes and granites. These rocks present surface water yields with marked seasonality of flow. Using purchased inputs to overcome the above land degradation scenarios in the traditional farming setting appears remote because farmers lack adequate cash and good input delivery networks.

The Bamenda Highlands constitute a unique geomorphological unit which coincides with the administrative unit known as the North West Province of Cameroon (Fig. 1) According to a 1999 United Nations Development Programme/Ministry of Planning and Regional Development (UNDP/MINPAT) report, the area is predominantly rural, with 86% of the population living in rural areas, and nearly two in three rural residents classed as poor. In terms of regional disparities, the 1996 household survey by the Ministry of Economy and Finance notes that a significant percentage of the poor live in the Northern Provinces of the country. The North West Province is ranked third out of the ten provinces with an estimation of 365,352 poor (Fonchingong, 2004). The World Bank report (1999) further notes that Women constitute 52% of the three million poor in Cameroon who cannot afford even the food components of a "consumption basket". This indicates a situation of extreme poverty (The World Bank, 1995). In Cameroon, women make up the majority of the poor trapped below the poverty line and poverty is particularly acute for women living in rural areas and heading households (UNDP, 1998). Women are therefore the victims of land degradation.

Fig. 1: Relief of the Bamenda Highlands

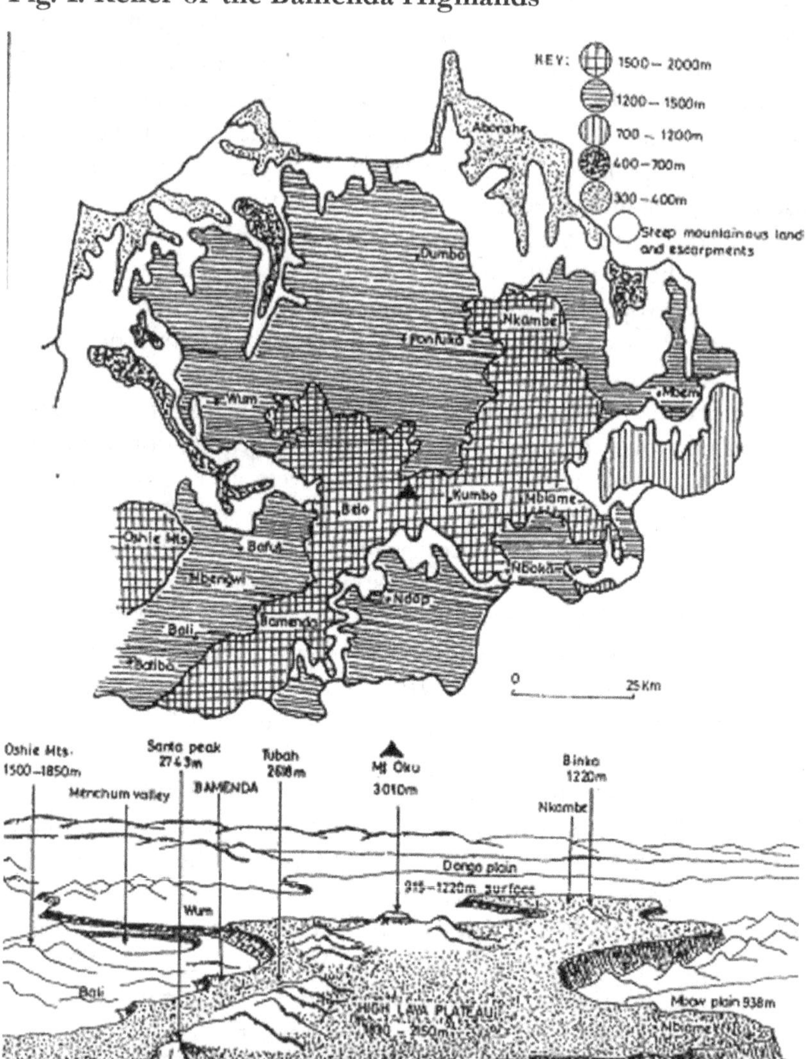

Poverty is rife in the study area as a result of high population density (over 96 inhabitants/sq. km) ruggedness of the terrain, remoteness from markets, the drudgery of transportation, isolation and the fragility of mountain environments. The North West Provincial Service for Statistics (1999) estimates an average farm family size of 10.2 persons, with the largest households having 13.2 persons. Most farm family heads (82%) are married and predominantly male (81%). In terms of literacy, 24% had no schooling, 54% primary 1 to 7 and 14% some secondary education. In rural areas 89% of households send children to school, that is, 71% of school age children from farming households attend school. As the foregoing suggests, the level of professional training in agriculture and general education in farm families is negligible. The Provincial Service for Statistics estimates it to be 2% and concludes that this notwithstanding, in 14% of farm households the agric-training level was as follows:

- 12% of male farm family heads received informal training in agriculture;
- 2% of female farm family heads received informed training in agriculture provided by various non-governmental organizations and the Agricultural Extension Service of the Ministry of Agriculture.

Agriculture is the backbone of the study area. A survey by the Provincial Service for Statistics (1999) indicated the following sources of farm family income: crop cultivation (90.8% of the farm families), livestock raising (52.5% of the farm families), fisheries (5.4% of the farm families), forestry (19.6% of the farm families), apiculture (16.2% of the farm families) and palm wine tapping (28.1% of the farm families). According to Fonchingong (2004), women constitute approximately 54% and men 46% of the economically active population.

Although publications in the human sciences are not directly concerned with the changes in the physical environment, Bawden and Landale-Brown (1961), Kaberry (1952), Keay (1959), Carter (1956), McCulloch (1948), Asombang (1983), Morin (1980), and Muller (1974), conclude that human agro-pastoral civilization in the Bamenda Highlands has increased the threat of environmental degradation. Studies by Masahiro (1984) show a dynamic relationship between the human impact and civilization under anthropogenic agents. It has been concluded that the wholesale destruction of the original forest of the highlands has reached the proportions of a major crisis (Dongmo 1984). Lightbody (1952) identified deforestation, fire and grazing as factors affecting the structure and composition of the Bamenda mountain grasslands. Lambi (1999 and 2001) expressed concerns about the degradation of land resources in the Bamenda Highlands.

Studies by Haruki (1984), Kadomura (1980), Kikuchi (1977), and Tamura (1982) attempted a reconstruction of the environmental history of the region. The conclusion from these studies is that extensive deforestation and land use intensification has resulted in surface wash processes, soil creep and slow down-slope movement of soil related to intensive continuous cultivation. Kyuma (1984) concludes that the savannization process is a consequence of this anthropic impact. Hawkin's and Brunt (1965) observed that soils in Bamenda Highlands show wide variations in terms of age, origin, physical and chemical properties. However, they have one thing in common: the strong dependence on organic matter for the maintenance of fertility. The rapid loss of fertility and of structure happens when organic matter is not added continuously in adequate quantities. Unfortunately, land use intensification and bush-burning inhibits the reconstitution of the climax vegetation, soil nutrients and soil structure.

Methods And Data Sources

The study used secondary data to realize an eco-floristic mapping of the area. The extent of land degradation was established using the vegetation map of 1965 (Hawkins and Brunt, 1965) and the vegetation map of 1987 (Hof and Kips, 1987). The mapping process was updated using field observations. This enabled a zoning of the area into eco-floristic regions, that is, lowland forest, savannah woodlands, montane forest and afro-alpine grasslands. The causes and effects of land degradation for each eco-floristic zone were investigated using field observations and informal interviews. Land resource potentials, land degradation factors and priority land resource management measures were established per ecological zone in qualitative terms.

Results And Discussions

Fig. 2 presents the land use intensity in the area. The main land use systems are extensive grazing systems with transhumance and crop cultivation. With a total land surface area of 17300 sq. km, the area has a population of 1,350,000 people and a density of 78 inhabitants /km^2 according to the 1987 population census. Demographic projections put the population at 2,090,300 inhabitants by 2010 (MINPAT/UNDP, 1999). This gives an average density of 120 inhabitants / sq. km. However, there are local pockets of high density areas in Mezam and Ngoketunjia with over 200 inhabitants /sq. km. According to the 1972 Agricultural census of the Ministry of Agriculture, the farm sizes ranged from 0.5 to 10 ha. The distribution of farm sizes was relatively skewed. Farms of less than 1.5 ha (72.7% of all farms) accounted for over 43.3% of the cultivated land. There are large and small farms. With demographic pressure on land, there is a tendency towards a reduction in fallow durations (2-3 years fallows) and extension of farms into marginal lands. Arable land accounts for about 45% of the total land surface while grazing land accounts for about 55%. About 5% of the land is unused (Fig. 2).

The average stocking rate is 3 head of cattle per hectare. The obstacle to cattle grazing is the shortage of forage and there is clear evidence of overgrazing in Santa, Sabga, Pinyin, Mvem, Abar, Esu, Tatum, Jakiri, Mbiame, Misaje and the Mbum plateau. Figure 3 and 4 present the vegetation of the area for 1965 and 1987 respectively. These maps enabled a division of the area into the main floristic zones (Table 1).

- *Lowland forest*: Sub-humid zone, low and medium altitude, semi-deciduous forest and woodlands.
- *Savannah woodland*: Sub-humid zone, medium altitude, variations, of savannah woodlands and mosaics of cropland/ gallery forests.
- *Montane Zones*: Altitude above 1 800m, montane forest and afro-alpine grassland.
- *Afro-Alpine Zones*: Mountain peaks, afro-alpine short grassland

Fig. 2: Land Use Intensity in the Bamenda Highlands

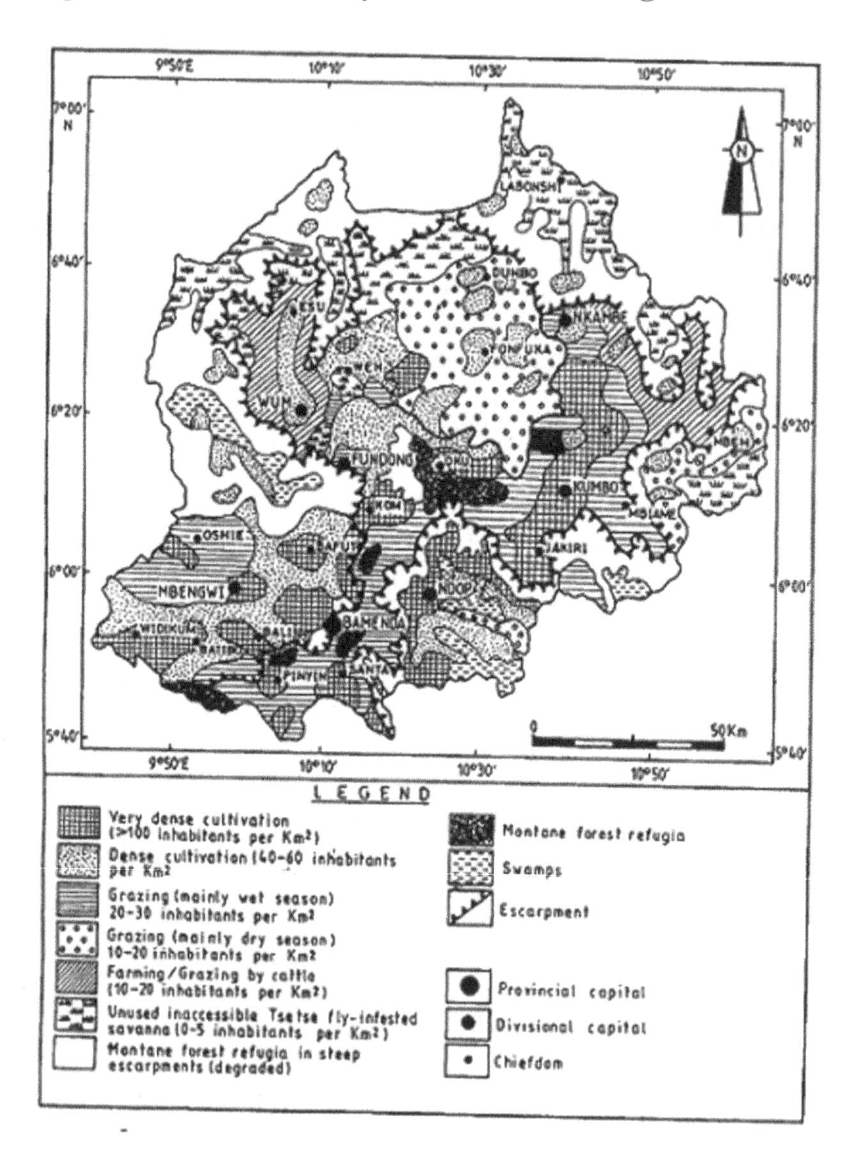

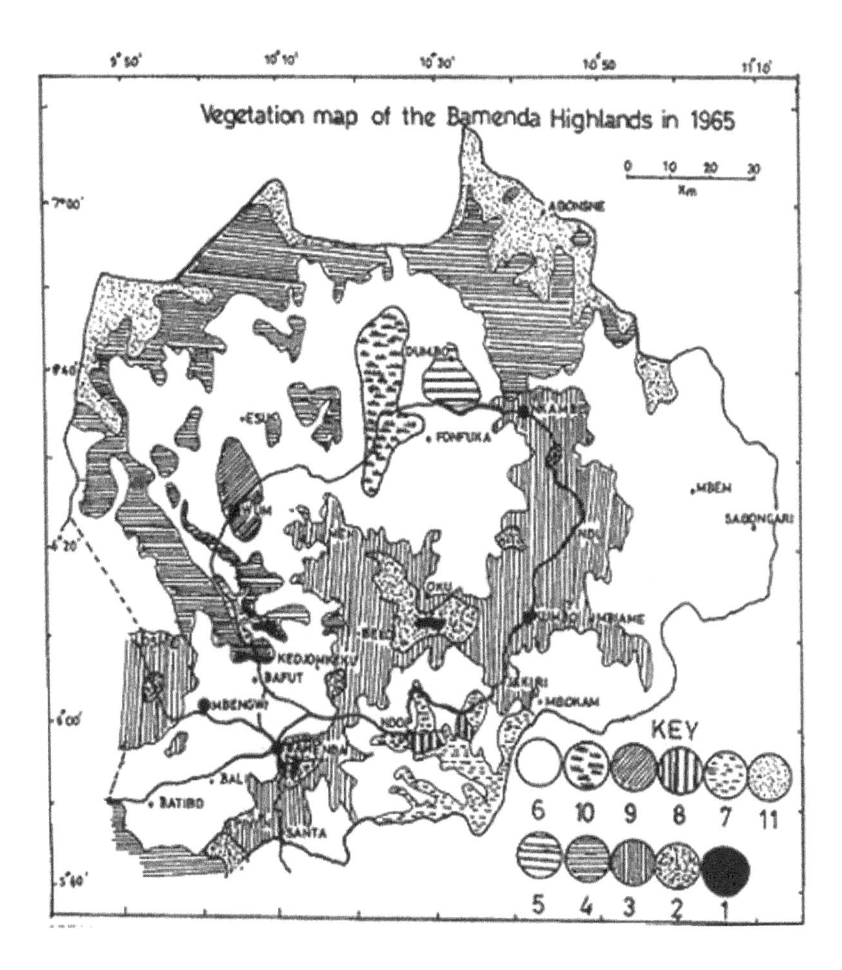

KEY: 1. Alpine bamboo forest or thicket 2. Moist montane forest 3. Hyparrhenia and Sporobulus grassland: derived from the moist montane forest 4. Moist evergreen forest 5. Annona-Nauclea tree and shrub savanna 6. Terminalia tree and shrub savanna 7. Seasonally flooded grassland 8. Swamp forest 9. Hyparrhenia and Beckeropsis grassland 10. Loudetia grassland 11. Southern guinea savanna.
* 5, 6, 9 and 10: Derived from moist evergreen forest

Table 1: Eco-floristic land units of the Bamenda Highlands

Climatic zones	Mapping units		Eco-floristic community
	Locations	Altitude	
Extremely hot and relatively wet zone	Donga plain, Kimbi, Katsina, Lower Menchum Valley	300m	Lowland forest
Hot, very humid and extremely wet zone	Batibo, Widikum, Mbembe and Mfunte	700m	Lowland forest
Hot and wet zone	Mbaw plain	800m	Savannah woodland
Warm and wet zone	Batibo, Kom-Bum area, Misaje, Wum, Esu, Bali, Mankon, Bafut plateau	900 to 1500m	Savannah woodland
Very hot and very sunny zone	Ndop plain	1150m	Savannah woodland
Cool and misty zones:	High lava plateau	1250 to 2250m	Montane forest
Cold, very cloudy and misty zone	Mountain peaks: Oku, Binka, Santa, Mendankwe	2250 to 3011m	Afro-alpine grassland

The main problems that arise from the exploitation of resources for each vegetation zone were established. The specific causes of the degradation of the environment and scope for sustainable land management were identified.

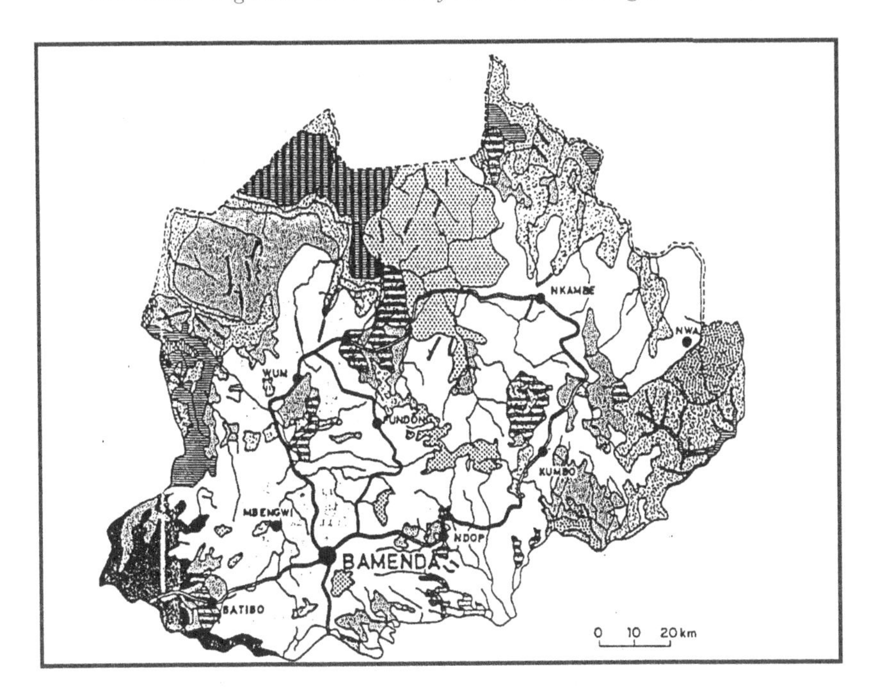

Woodland savanna (*Burkea africana, Daniellia Oliveri, Borassus aethiopum*).

Tree savanna and shrub savanna (*Daniellia Oliveri, Lophira lanceolata*) generally with a dense network of gallery forest

Grassland.

Montane zone (altitude above 1800m, rainfall exceeding 1500mm)

Evergreen mountain forest (trees generally short; composition variable according to locality; *Podocarpus milanjianus* common).

LLANEOUS

Mosaic of cropland, grassland and savannas

Lake Bamendjing

Gallery forest

1. Savannah woodland

Hunters burn to flush animals out of their natural niches and to expose them to the reach of fatal weapons. Sometimes, when most of the hills and highlands are burnt, the animals get concentrated in the gallery forest, water-courses and river beds where they can easily be surrounded, slaughtered and collected. Burning to improve farm yields is done mostly for cultivation. Stockmen set fire to renew grasses and herbage and sometimes to kill certain disease vectors like ticks (Figure 5). The net result of these fires is the destruction of the biomass. This is further strengthened by the fact that once the fire is set, the author does not consider how far it can spread and what and how much the fire can damage. The devastating effect of destruction on the environment comes from overgrazing and/or overstocking pasture land. The savannah woodlands occupy intermediary relief and it is here that most water catchments are found. Following the destructive actions of uncontrolled bush-fires and pasture degradation, water catchments are destroyed. The precursors or contributing elements to the destruction of water catchments are unauthorized grazing, cultivation, burning of vegetation and planting of unsuitable species of trees and vegetation. Soil erosion, the most serious agent of denudation, contributes enormously to the degradation of the environment in the Savannah woodlands. It results from situations which expose the soil such as bad civil engineering activities and bad farming practices such as ridging down the slopes.

The irrational exploitation of the forest constitutes a serious menace. This irrational exploitation leads to deforestation, which in turn degrades the environment. Deforestation also results from invasion by farmers in search of fertile land which in effect results from the accumulation of biomass in the form of dead leaves over the years. Deforestation by farmers is a consequence of shifting cultivation (Fig. 6). Faunal and phyto-genetic reserves form part of the environment of the Savannah woodland zone. Encroachment into the zone by the population for diverse reasons has been a major environmental concern. Rare species are exterminated with detrimental consequences for biodiversity and the tourism industry. The encroachment takes various forms: poaching, illegal settlement or squatting, illegal farming, and grazing by agropastoralists.

To reduce and attenuate the effects of uncontrolled bush-fires, a policy should be designed and enforced on the control of bush-fires. This policy should be developed with the participation of grass roots communities and should include raising the awareness of the population sufficiently. Techniques to improve the management of pastures need to be implemented. Other inputs like improved forage seeds and planting material and techniques for optimal stocking rates can check pasture degradation.

Savannah woodland is the ideal zone for harnessing water catchments. To protect these water catchments, certain actions need to be carried out. Among these one can cite the creation of community-based water catchments management committees. These committees will oversee operations such as replanting catchments with suitable vegetative species. Soil erosion, an obstacle to the attainment of a rationally managed environment, can itself be reduced by the enforcement of techniques such as agro-forestry, improved fallow, contour farming and the enforcement of improved methods of soil conservation.

Fig. 5: Model of the Shifting Cultivation Cycle and Degradation of Forests

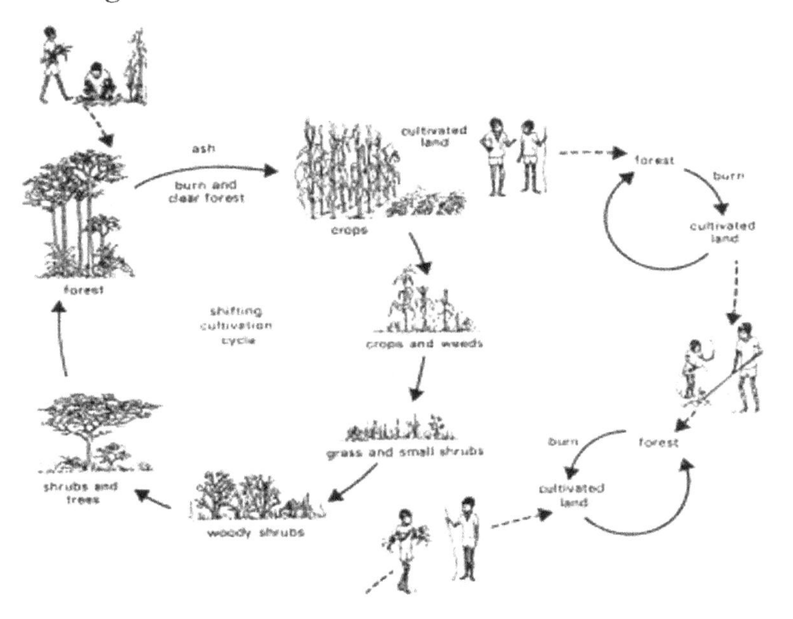

The principal resources of the savannah woodland are trees and grasses. The question of the rational exploitation of forests requires special attention. To exploit the forest rationally, certain conditions have to be met and certain operations have to be undertaken. Some of these are the institution of appropriate management plans and coordinating forest management activities as well as reforestation of degraded forests with suitable trees species. Reserves constitute biodiversity sanctuaries whose *raison-d'être* is the preservation and the enhancement of the environment. To reduce encroachment into these reserves, some measures have to be taken, such as the elaboration of a management plan, proper studies, and the demarcation of reserves.

Fig. 6: Model of grazing impact on forest-savannah contact zones and the process of savannization: D = dry season, R = wet season (modified after Hori, 1986)

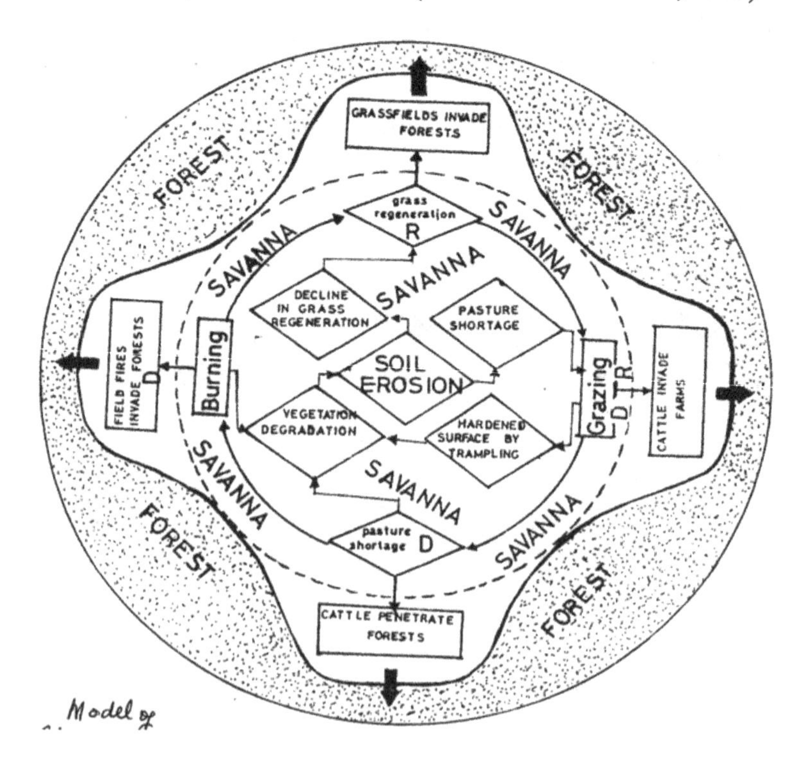

2. Lowland forest zone

The deforestation of the lowland forest is closely linked to inadequacies in the supply of forest products to the rural masses. In the absence of alternative low-cost energy sources, there is a high dependence on the forest for fuel wood. Not only do they depend on the forest for fuel wood but the remnants of these forests also constitute the principal supply of medicinal plants and all forms of timber for the construction of houses, bridges and other related domestic and commercial uses. Similarly, as most lands are taken over for arable cropping, cattle owners increasingly invade these patches of forest for grazing, and by so doing, regularly burn to encourage forage re-growth for stocks. Hunters and honey collectors also set the forest on fire to force out animals during hunting and honey collection respectively. Apart from poverty and the absence of alternative sources of animal protein, the lowland forest is also irrationally exploited as a result of ignorance. Therefore deforestation constitutes an important cause of environmental degradation. The main reasons for hunting are to provide animal protein (bush meat) and feathers, horns, skin, bones and other animal products for socio-cultural purposes. Poaching to secure feathers, skins and animal skulls is important. Similarly, hunting for snakes has become a profitable commercial activity in the zone and bush meat (game) markets are developing along roads and in urban centres. These activities contribute to biodiversity degradation.

In order to reduce deforestation in the zone, a number of inter-related actions are necessary. There is first of all, a need to carry out a complete inventory in order to identify areas where research and/or conservation efforts need to be concentrated. At community level, it would be important to carry out an awareness-raising campaign so that inhabitants of the forest areas and particularly those who depend on it for livelihood begin to appreciate the tangible and intangible benefits of forests. Through this action, they may abstain from activities that destroy the forest. The boundaries of most forest areas are vaguely defined and it becomes difficult to design efficient management plans. There would therefore be a need to demarcate forest areas from grazing and farming land. Because forestry and agroforestry research has concentrated on exotic species, there is a need to emphasize research on the propagation methods of potentially useful indigenous tree species for which silvicultural knowledge is either lacking or inadequate.

The implementation of the above measures without affordable alternative energy sources would hardly be successful. Therefore, as complementary measures, there is a need to promote the large-scale production of energy-efficient firewood stoves while research should also be carried out to identify alternative energy saving sources. There is also a need to intensify research and extension efforts in the use of improved farming practices in order to increase yields per unit area of land. This can help to reduce encroachments into the forest for cultivation.

The intensification of anti-poaching controls can be successful if affordable alternative sources of meat are available. The promotion of small-scale livestock production, particularly rabbit and fish, should be encouraged. Appropriate methods must be determined to involve the population in anti-poaching activities as well as in the overall management of biodiversity.

3. Montane forest zone

The deforestation of the montane forest is closely linked to inadequacies in the supply of forest products to the rural masses. In the absence of alternative low-cost energy sources for the rural masses, there is a high dependence on the forest for fuel wood. The remnants of these forests constitute the principal supply of medicinal plants and all forms of timber for the construction of houses, bridges and related domestic and commercial purposes. Similarly, as most lands are taken over for arable cropping, cattle owners increasingly use these patches of forest for grazing, and so regularly burn them to encourage forage regrowth.

It has already been pointed out that hunters and honey collectors also set the forests on fire to flush animals during hunting and honey collection respectively. Apart from poverty and the absence of alternative sources of animal protein, montane forest is also irrationally exploited due to the ignorance. Deforestation constitutes an important cause of environmental degradation. Hunting has also been seen as an occupation (Macleod, 1986; Stuart, 1986). These activities contribute to biodiversity degradation. Figure 7 presents a model of the degradation of montane forests in the area. Periodic burning, felling and grazing initiate a savannization process.

The deforestation of water catchments primarily results from afforestation efforts with unsuitable tree species particularly the Eucalyptus. The tree has a very deep tap root and extensive rooting system requiring huge quantities of water for growth. Uncontrolled bushfires around water catchments also constitutes a serious constraint necessitating the establishment of fire breaks.

Soil erosion is the most serious agent of denudation in the zone and is responsible for the extensive breakdown of the fragile soils. Overgrazing, which leaves the land bare, together with farming down slopes as well as the slash, burn and bury method of cultivation, enhance soil run-off and the breakdown of soil structures. Heavy civil engineering activities especially along road also increase the potential of erosion. Furthermore, the planting of unsuitable tree species, particularly eucalyptus, quickly dries up the soil and when the vegetation is burnt down during the dry season by farmers and graziers, erosion is again facilitated.

4. Montane grasslands

In the Montane grasslands of this zone, *Hyparrhenia* dominance (an ungrazed fire climax) gives way to *Sporobolus* dominance (a grazed fire climax). The final result of the use of uncontrolled bush fires, trampling by cattle and continuous overgrazing of the pastures is the invasion and spread of bracken fern (*Pteridium aquilinum*) which further suppresses the growth of other palatable pasture species.

Overgrazing, results from overstocking, having as underlying causes demographic expansion and pressure on the land, increase in cropping land area and the consequent reduction of rangeland areas, shifting cultivation, bush invasion, increase in undesirable pasture species, and unrestricted ownership of livestock kept for prestige. The range management practices that cause pasture destruction in the zone are mainly the traditional practices which include overgrazing, overstocking and the use of uncontrolled bushfires.

Environmental degradation also results from ridging down slopes, torrential rains, trampling by cattle, topography, and destruction by livestock and civil engineering activities. The hilly nature of the zone renders it susceptible to high run-off and this is accelerated when ridges are constructed down slopes. Overstocking leads to trampling of the soil, causing hard pans and increasing run off and erosion.

Fig. 7: Model of the Dynamic Relationship between Human Impact and the Process of Savannization

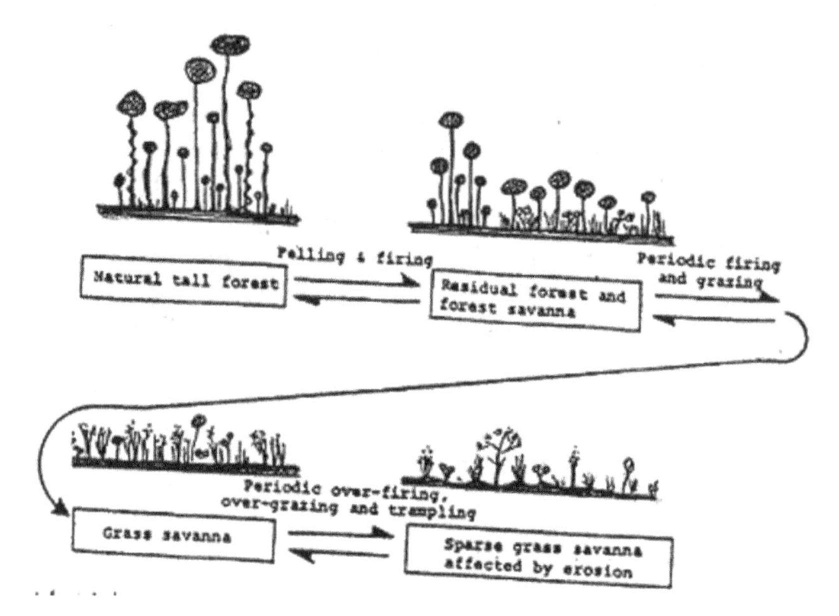

The unique fauna and flora of the zone makes it endemic for some species. However, illegal hunting, ignorance on the part of the population of the importance of rare species, uncontrolled use of chemicals, overgrazing, and excessive exploitation of medicinal plants destroys biodiversity. The Afro-Alpine zone is also a major watershed and various activities of man in the area have negative impacts on the watershed. These activities include: inappropriate farming practices, deforestation, overgrazing, uncontrolled use of chemicals and afforestation with unsuitable tree species. Farming practices that encourage soil erosion and run-off prevent infiltration, and as such, tend to have negative impacts on the watershed. Overgrazing, like deforestation, lays bare the soil and encourages erosion. The uncontrolled use of industrial chemicals for de-ticking cattle and the use of fertilizers lead to pollution of the watershed. Afforestation of the watershed with unsuitable tree species like eucalyptus also tends to dry up the watershed.

Fig. 8: Model of Poor Natural Resource Management in the Bamenda Highlands and Strategies for Sustainable Resource Management

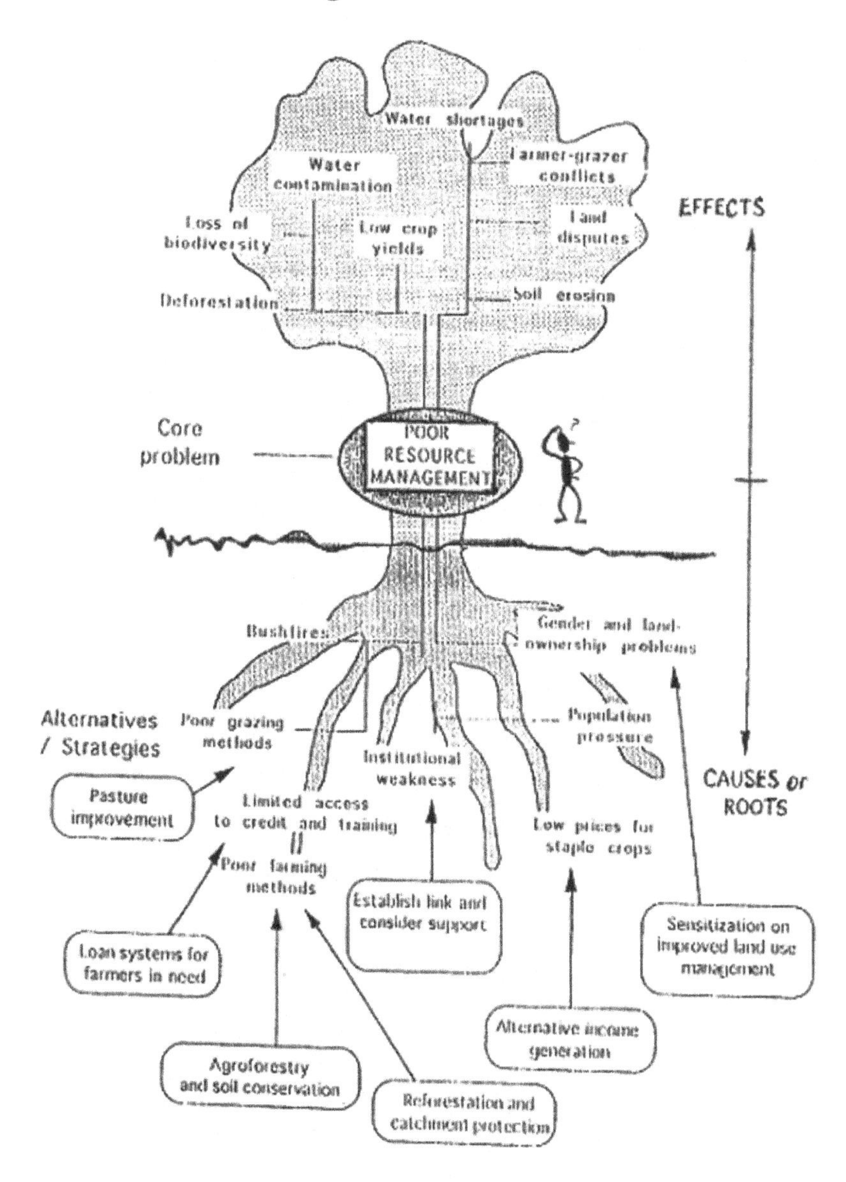

Figure 8 summarizes the core problem, its causes and its effects. Land degradation is caused by gender and land ownership problems, population pressure, poor grazing techniques, poor farming methods, low prices for staple foods, institutional weakness, bush fires, and limited access to credit and training. The main effects of land degradation are water shortages, farmer-grazier conflicts, land disputes, water contamination, soil erosion, low crop yields, loss of biodiversity and savannization (deforestation). Sustainable land management strategies in the model (Fig. 8) include pasture improvement, promoting agroforestry and soil conservation, raising awareness on improved land management, promoting alternative income generation activities, institutional and human capacity building in land resource management, facilitating rural people's access to credit and training, and promoting reforestation and catchments protection projects (Table 2).

Table 2: Reforestation and forest conservation projects and related land use conflicts in the Bamenda Highlands

Location	Land Use Conflict with Communities	Type of Action Required
Mezam	- Bali-Ngemba (state)-(farming, grazing and illegal exploitation)	- Afforestation to be continued - Regeneration
	- Bafut-Ngemba Forest(s) farming, grazing and illegal exploitation	- Afforestation to be continued - Regeneration
	- Bafungi Forest (communal) farming, grazing and illegal felling of trees	- Regeneration
	- Bafut-Tingo valley forest (communal farming and exploitation	- Regeneration
	- Bambui-Mendankwe forest range (communal) grazing, farming and exploitation.	- Afforestation - Regeneration
Ngoketunjia	- Ndop Plain gallery forest (c) (farming and exploitation)	- Afforestation
	- Bafanji forest (c) (farming and exploitation activities)	- Regeneration

Table 2: Reforestation and forest conservation projects and related land use conflicts in the Bamenda Highlands (Contd)

Location	Land Use Conflict with Communities	Type of Action Required
Momo	- Widikum/Menka forest (farming activities)	- Regeneration
	- Oshie/Ekwere forest (c) (farming and grazing acxtivities)	- Regeneration
	- Gwofon-Oshum forest (c) (farming and exploitation activities)	- Regeneration
	- Ngemba /Mbengwi (council) (farming and exploitation activities - Acha/Tugi forest (c) farming, grazing and exploitation activities)	- Afforestation - Afforestation - Regeneration
Boyo	- Su-Bum forest (c) (grazing, farming and exploitation	- Regeneration
	- Mbueni-Menjang forest (farming, and exploitation	- Regeneration
	- Ijim Mountain Forest (farming, grazing, and exploitation	- Ongoing project needs monitoring and strengthening
Bui	- Mbokam forest (c) (farming and grazing	- Regeneration
	- Nkobifen forest (c) (farming and exploitation)	- Afforestation - Regeneration
	- Kilum Mountain Forest (to be constituted) (grazing, farming and exploitation)	- Ongoing project needs monitoring and strengthening
	- Bui-Mbim forest (c) (farming and exploitation)	- Regeneration
	- Kuflu forest (c)(farming and exploitation)	- Afforestation - Regeneration

Table 2: Reforestation and forest conservation projects and related land use conflicts in the Bamenda Highlands (Contd)

Location	Land Use Conflict with Communities	Type of Action Required
Donga/ Mantung	- Mbak forest (c)	- Regeneration
	- Berabe/Mbembe forest (c) (farming and exploitation)	- Regeneration
	- NjisinglTabenken forest (c) (farming, grazing and exploitation)	- Regeneration
	- Mbfu/Tabenken forest (c) (Farming, grazing and exploitation)	- Reconversion using native species
	- Nkanchi Forest (c) (Grazing, Farming and exploitation)	- Regeneration - Afforestation
	- Masim valley Forest (c) (Farming and exploitation)	- Regeneration
	- Mbibi/Tala Forest (c) (Grazing, Farming and exploitation)	- Regeneration
Menchum	- Menchum Valley Forest (c) (Farming and exploitation)	- Regeneration
	- Kom/Wum Forest(s) (Farming and exploitation activities)	- Afforestation - Regeneration
	- Fungom Forest Reserves (s) (Farming and exploitation)	- Regeneration
	- Furu-Awa Forest (c) (indiscriminate logging by Nigerians)	- Protection and creation of access roads.

Key: S=state forest or forest reserve, C= Communal forest

Conclusion

The study analyzed the problems of land degradation and its effects on rural livelihoods. Poverty and the need to sustain livelihoods are the principal cause of land resource degradation. The degradation of the environment due to poor resource management is the core problem. This is illustrated by the following:

- Encroachment into reserves, uncontrolled bushfires, soil erosion, deforestation, pastures destruction and water catchments destruction in the savannah woodland.

- Deforestation and poaching for the lowland forest.
- Deforestation, poaching and soil erosion in the montane forest.
- Uncontrolled bush-fires, destruction of pastures, soil erosion, biogenetic erosion, and destruction of watersheds in the afro-alpine grassland.

The immediate effects of these land degradation scenarios on the livelihoods of rural people are:

- *Reduced farm and grazing land*: The result is the cultivation of distant farmlands and the perpetuation of transhumance. Because most distant farms are fallow lands where cattle graze, farmer/farmer and farmer/grazier confrontation lead to poverty, misery, loss of human dignity and increased death rates.
- *Declining soil fertility and quality of rangeland*: results in low animal and crop yields, lead to low income, misery and therefore human disaster.
- *Contamination of water sources:* also results in disease prevalence, rampant health hazards, misery and increased death rates.
- *Reduced spring flows:* As a result of the constant deforestation around crater lakes and water catchments areas, the regularity of spring flow as well as the intensity of water flows are substantially reduced.
- *Loss of biodiversity:* As the natural habitats of the various plants and animal life are destroyed, the outcome is the continuous loss of biodiversity as well as tourist potential.
- *Migration:* This is a direct consequence of the depletion of the natural resource base, which pushes people to other places where more resources can be found.

There is therefore an urgent need to develop appropriate sustainable land resource management strategies. for this mountain region of Cameroon.

References

Asombang. R. B. (1983): Report on Archaeological Research at Mbi Crater and Shum Laka in North West Province. D.G.R.S.T. Yaounde, 17 pp.

Bawden, M.G. and Langdale-Brown, I. (1961) An Aerial Photographic Reconnaisance of the Present and Possible Land in the Bamenda Area. Southern Cameroon Dept. Of Tech., Coop., D.O. S. Forestry and Land use Section. 25 pp.

Carter, J. (1956) The Fulani, Their Cattle and the Grazing Lands of Bamenda Province of Southern Cameroon, unpublished manuscript.

Denniston, D. (1995) Sustaining mountain people and environments. *in* State of the world 1995; A Worldwatch Institute report on progress towards a sustainable society. W.W. Norton and company, New York and London, pp 38-57.

Dongmo, J. L. (1984) *Le Rôle de l'Homme à Travers ses Activités Agricoles et Pastorales L'Ouest Cameroun*,in: Kadomura H (ed.), Natural and Man-induced Environmental changes in Tropical Africa, Hokkaido Univ. Sapporo, pp 61, 74.

Fonchingong, C. (2004) Integrating gender concerns for livelihood improvement and local development in North West Cameron: The case of NGOs, in: Journal of Applied Social Sciences, University of Buea.

Haruki, M.(1984) Silviecological studies in the forest zones of Cameroon. In: H. Kadomura (ed.), Natural and man-induced environmental changes in tropical Africa: case studies in Cameroon and Kenya, Hokkaido University, Sapporo, pp 75-91.

Hawkins, P. and Brunt, M. (1965) Soils and Ecology of West Cameroon. Report No. 2083 Rome F.A.O.

Hof, J and Kips, P. (1987) Land Evaluation: General Metholodology and Results for the Ring Road Area, with Emphasis on low Input Maize, Smallholder Coffee, Oil Palm and Extensive Grazing, FAO/UNDP Soil Resource Project.

Hori, N. (1986) Man-induced landscape in a forest-Savanna contact area of East Cameroon, in: H. Kadomura (ed.), Geomorphology and environmental changes in tropical Africa: case studies in Cameroon and Kenya. Hokkaido University, Sapporo, pp. 45-62.

Kaberry, P.M. (1952) Women of the Grassfields: a study of the economic position of women in Bamenda, British Cameroons. London, Colonial Research Publications, pp.14.

Kadomura H. (1980) Reconstruction of Palaeo-environments in Tropical Africa during the Wurm Glacial Maximum: A Review. Environ. Sci. Dept. University of Hokkaido, Japan. Vol. 3. No. 2, pp 147-154.

Keay, R.W.(1959) Montane Vegetation and Flora in the British Cameroons. Proc. Linn Soc. Lond., 165 p.

Khan, A. (2002) Comments on poverty and sustainable Livelihood. BGMS-82 Mountain Forum Discussion. Ar 20/03/2004. Mountain Forum Atlas Research results: Eastern Arc.

Kikuchi, T. (1977) Volcanic Landforms of West Cameroon, in: Kadomura (ed.), Geomorphological Studies in the Forest and Savanna Areas of Cameroon. Publication No. 1. Univ. of Hokkaido, pp 53-60.

Kyuma, R. (1984) The ecologyof shifting cultivation. Scientific American 14(4), pp 20-31.

Lambi C.M. (1999) Land degradation in the North West Province of Cameroon. Culture and Environment, J. Dunlop and W. Roy (eds.) University of Stratchclyde /University of Buea, Cameroon, pp 174 -183.

Lambi, C.M. (2001) Environmental issues: problems and prospects. Unique Printers, Bamenda, pp 45-66.

Lightbody, J.B. (1952) The Mountain Grassland Forest of Bamenda and Factors influencing their structure and composition. Commonwealth Forest Institute, Oxford.

Macleod, H. (1986) The conservation of Oku Mountain Forest, Cameroon. Cambridge, International Council for Bird Preservation. Study report No. 15. pp 53-58.

Masaharu Haruki (1984) Late Quatenary Environmental Changes in Southern Cameroon: A Synthesis, in Kadomura H. (ed.) Geomorphology and Environmental Changes in Tropical Africa: Case Studies in Cameroon and Kenya. Hokkaido University, p. 153.

McCulloch J. (1948) Grazing Improvements in the Bamenda Division. Cameroon Under British Mandate. *Bull. Agric. Du Congo Belge*, Vol. 40.

Ministry of Agriculture (1972) Agricultural census for 1971/72, Ministry of Agriculture, Yaounde.

Moby, E. (1979) Climate, in Louange, J-F. (ed). Atlas of the United Republic of Cameroon. Editions Jeune Afrique, Paris, pp 16-19.

Morin S. (1980). *Apport des Images LANDSAT à la Connaissance de la structure des Hautes Terres de 1 'Ouest Cameroun*, in: Cameroon Geographical Review 1. pp 181 -196.

Muller J. P. (1974) *Aptitude Culturales des sols de l'Ouest Cameroun. Notion d'Etablissement et Utilisation des cartes*, ORSTOM, Yaounde.

Nkwi, P.N. and Warnier, J-P. (1982) Elements for a history of the Western Grassfields. Dept. of Sociology, University of Yaounde. 236 pp.

Provincial Service for Statistics (1999) Baseline Survey of the rural world situation in the North West Province of Cameroon, Provincial service of Statistics NWP, Bamenda.

Tamura, T. (1986) Regolith stratigraphic study of late Quaternary environmental History in West Cameroon Highlands and Adamawa, in: H. Kadomura (ed.), Geomorphology and environmental changes in the forest and savannah Cameroon. Hakkaido University, Sapporo.

Tamura, T. (1982) Recent Morphogenetic Changes as Revealed in the Slope form and Regolith Characteristics of the West Cameroon highlands, in: H. Kadomura (ed.), Geomorphology and Environmental changes in the forest and Savanna Cameroon. Hakkaido Univ., Sapporo, p. 67—78.

Stuart, S.N. (1986) Conservation of Cameroon Mountain Forests. Report of the ICBP Carneroon Mountain Forest Survey. ICBP.

The World Bank (1995) Cameroon: Poverty assessment. The World Bank, Washington D.C.

The World Bank (1999) A 1999 update of the Cameroon poverty profile. The World Bank office, Yaounde, Cameroon

UNDP (1998) Integrating human rights with sustainable development: a UNDP policy document, UNDP, New York.

UNDP/MINPAT (1999) Regional socio-economic study of Cameroon. The North West Province Project, UNDP office, Yaounde.

Yamoah, C.; Ngueguim, M.; Ngong, C. and Cherry, S. (1994) Soil fertility conservation for sustainable crop production: Experiences from some highlands areas of North West Province, in: Proceedings of Agroforestry Harmonization Workshop. GTZ, USAID, and Helvetas, RCA Bambili, pp 1-6.

Chapter Three

Implications of Rapid Urbanisation for Floods, Sediment and Debris Flow Hazards in Bamenda, Cameroon

Summary

In developing countries, rapid urbanization induced by rural poverty forces recent migrants from the countryside frequently to settle in slums that carpet steep hillsides and flood plain zones. These spreading slums are prone to catastrophic flood, landslide, sediment and woody debris hazards. Understanding the characteristics of these geomorphic and fluvial processes is indispensable for establishing effective measures to avoid and to mitigate the damages. The paper clarifies their characteristics and mechanisms in terms of trigger factors, processes, magnitude of floods, sources and yield of sediments and debris. It uses a combination of field measurements, observations and secondary data to establish these characteristics. It concludes that the lack of understanding of the mechanisms of these processes and the land use systems of the urban poor exacerbate these disasters. They are triggered by a combination of torrential rain, loose material of weathered rock, high discharge of woody debris, urban solid wastes in streams and flood-prone zones and rapid housing developments in disaster-prone areas. This social condition exacerbates the disasters. The paper therefore emphasizes the role of decision-making on land use planning and the scope for designing appropriate solutions to mitigate the hazards.

Introduction

Population experts forecast that by the year 2000, more than half of humankind would reside in huge metropolitan regions (Dolgoff, 1990). According to them, the most spectacular growth will occur in developing countries, where recent migrants from the countryside frequently settle in shanty towns. They settle on steep hillsides surrounding older central cities (Tyler, 1982) and other risk-prone zones. According to Wijkman and Timberlake (1984) floods affected 5.2 million people a year in the 1960s compared with 15.4 million in the 1970s – an almost three-fold increase. Between 1964 and 1982, floods killed 80.000 people and affected 221 million worldwide. In the major cities of developing countries, between 30 and 70% of the urban population live outside the law on steep slopes and swampy ground prone to flooding. As the poor spread across swampy zones and mudflats, the rest of the city spreads up steep slopes (Lambi, 1989; Lambi and Fogwe, 2001; Tadonki, 1999; Mainet, 1978).

In recent years, urban floods and associated landslides have hit the headlines of newspaper tabloids and news headlines in Cameroon. During the last decade, the cities of Maroua, Yaounde, Kribi, Douala, Limbe and Bamenda have experienced catastrophic flood events. On the 9th of August 2000, floods destroyed homes in the low-lying areas of Bamenda city. These were accompanied by catastrophic landslides which buried homes and squatters in the mass of earth which wasted from the slopes. Subsequent events took place in 2002, 2003, 2005 and 2006. Each event compelled municipal officials to convene emergency meetings which often spared no rhetoric to get things right. Yet, things always go the wrong way as manesfested by the recurrent catastrophes. This situation is due to the paucity of studies on the effects of poverty-induced urbanization on floods, slope stability, sediment and debris generation (Yasuhiro, 2001). Furthermore, the characteristics of these disasters are yet unknown. This study seeks to clarify the characteristics of urban floods, sediment and debris sources and yields and their urban management implications as they relate to vulnerable social groups.

Study Area

Bamenda is the provincial capital of the North West Province of Cameroon. In 2005, its population was 427,147 inhabitants and it is projected that this will rise to 907,766 inhabitants in 2015. Rapid urbanization and the ribonning growth of the urban field have resulted in the sprawling of householdings in flood plains and steep slopes. The town is located on a defensive site at the foot of an escarpment. Major landform units are steep escarpment slopes in upland areas, lower shoulder levels and an undulating, extensive, granitic foothill zone occupied by the town (479 hectares).

Streams lack clearly defined head streams and reach the foothill zone through small cascades where they form extensive flood plain zones. These shallow valleys have a climax vegetation of *Raphia Vinifera* swamp forests while the uplands are characterised by sub-montane forests. These floristic communities have largely been degraded by urbanization (Ndoh-Nwi, 1996).

The rainy season lasts from mid-March to mid-November. The rest of the year is dry. Mean annual rainfall is 2,824mm and average temperatures are 24°C in the undulating foothills and 20°C in the upland areas (Hawkins and Brunt, 1965). From August to October, catastrophic floods and landslides occur. Municipal authorities estimate that about 40% of the population (170,858) live in breach of the urban planning and housing regulations and generate garbage which is not accessible to municipal waste management services. The garbage is discharged in streams, flood plain zones, gullies, landslide scars and drainage reticulation systems (T'Hart, 1995).

The frequency of catastrophic flood and landslide events has increased since 1998. These often cause damage to houses, structures, disrupt communication and lead to the loss of human life and property. The escarpment zone and flood plains are typical examples of human colonization of hazard zones without an awareness of the risks involved. Human actions in flood-prone and landslide-prone zones exacerbate these hazards (Chefor, 2000).

Research Methods

The study investigated three main components of torrents. These were floods, slope instability, sediment and debris sources and their yields. Seven locations vulnerable to catastrophic flood peaks were monitored. With the assistance of base maps and aerial photographs produced by the National Geographic Institute in Yaounde, these locations were mapped. Quantitative data on peak discharges causing floods was obtained from the seven locations using Manning's Equation. The procedure is thus: the extent of flood marks left by swollen streams was investigated in the seven flood plain valley walls. Using the flood marks by the streams, the cross-sectional area of the streams at the height of flood was calculated. The degree of slope downstream then provided the other component of discharge velocity. By using the equation known as Manning peak discharge, (Q) was estimated (Malcolm 1983) as follows:

$$Q = \frac{A \times R^{1.5}\ S^{0.5}}{\eta}$$

where: A = the cross-sectional area of the channel at peak flow, S = slope as a decimal, η = Manning's roughness coefficient (this coefficient accounts for the frictional resistance to flow. It is around 0.030 for lowland streams and 0.045 for rocky upland streams), R = the hydraulic radius of channel at peak flow:

$$R = \frac{A}{W + 2d}$$

where: W = channel width and d = channel depth. Quantitative data so obtained yielded the magnitude of peak floods.

The rainfall intensity is a natural trigger of floods, landslides, sediment and debris flow. Due to the absence of automatic recording rain-guages (autographic gauges) in the study area, the rainfall intensity was estimated by establishing the duration of fall of the highest rainfall in minutes using a stop watch and the corresponding rainfall volume using a rain gauge. This was established for August to October (period with flood events).

Landslides and associated sediment and plant debris were also investigated. The objective was to establish the geomorphic and fluvial processes involved and the sources and yield of debris and sediment. Field observations were complemented by data obtained from Lambi (2004, Eze and Ndenecho (2004) and Ndenecho (2006).

Using a combination of field observations and documented data by Acho-Chi (1998) the hydrological conditions of the riparian zones in the town and the anthropogenic factors exacerbating the hazards were established by estimating the rate of urbanization and its impact on flood and landslide incidences, debris and sediment generation.

Presentation of Results and Discussion

Table 1 presents data on urban growth in Bamenda, that is, the process of increasing the percentage of the population living in the city. The population increased more than four-fold between 1976 and 1987.

Table 1: The growth in the population of Bamenda

Year	Total Population	General Increase	Percentage Increase	Annual increase (%)
1953*	9,765	-	-	-
1964+	18,489	8,724	98.3	8.1
1965+	19,000	1,489	2.7	2.8
1968+	25,900	6,900	36.3	12.1
1970+	33,376	7,476	28.9	14.1
1976*	44,764	11,388	34.1	5.7
1979+	52,537	7,773	17.4	5.8
1987*	203,480	150,943	287.3	35.9
1992+	257,200	53,720	26.4	5.3
1993+	270,400	13,200	5.1	5.1
2005+	427,149	156,749	57.9	4.8
2015+	907,766	480,617	112.5	11.2

Source: * *national census data*

+ Figures interpolated by the Provincial Service for Statistics, Bamenda

It is also projected that the population will more than double between 2005 and 2015. Uncontrolled population growth and diminishing job opportunities in overpopulated rural areas force people to migrate from the countryside into the city. There, too, jobs are seldom available. For the rapidly growing number of migrants, the city becomes a poverty trap. These urban poor are forced to live in crowded flood and landslide-prone areas (Fig. 1) where they face poverty (grubbing for food and fuel wood, minimal housing, open sewers, and untreated drinking water from filthy sources).

Inspite of landslide and flood hazards, joblessness and squalor, these slum dwellers cling to life with resourcefulness, tenacity and hope. Land value in these precarious sites is cheap and affordable. On the balance, most tend to have more opportunities and are often better off than the rural poor they left behind. They therefore acquire householdings in these precarious areas where they exacerbate hazardous fluvial and geomorphic processes in breach of land use zoning and town planning regulations. (Fig. 1 and plate 1).

Fig. 1: Seasonally Flooded Zones In Bamenda. Figure 2 Presents a Typical Cross Section of Flood-Prone Zones (Source: 2005 Fieldwork)

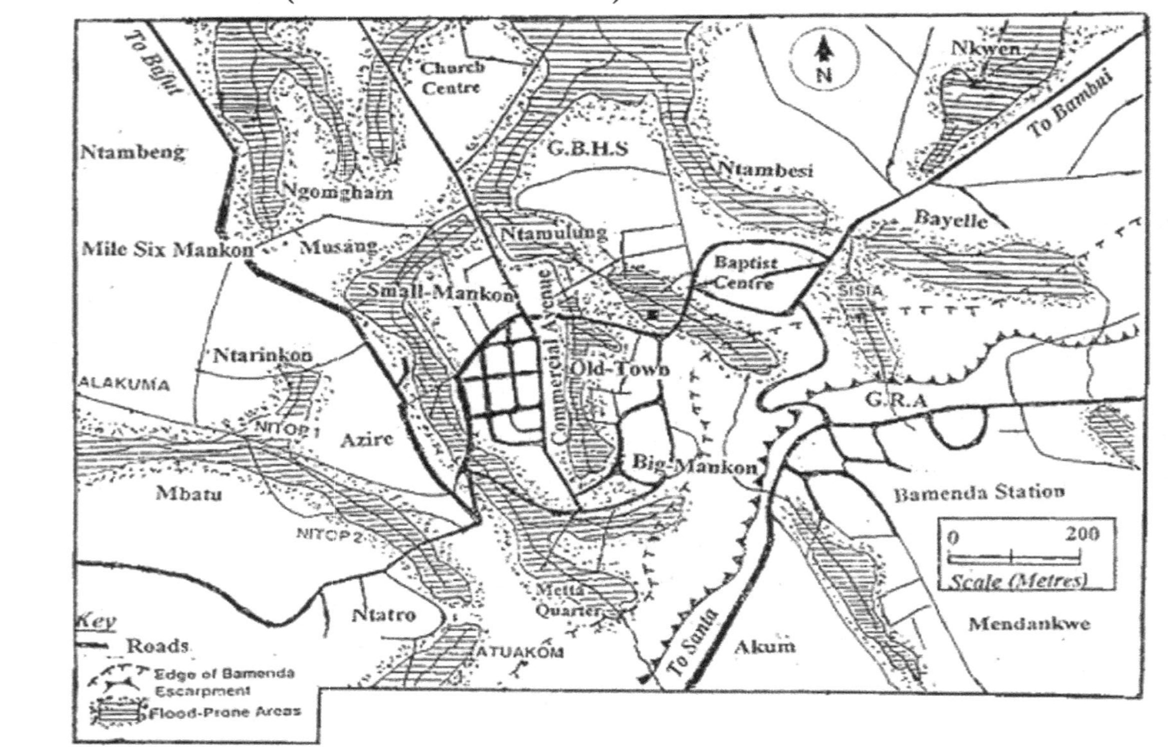

Plate 1: The Bamenda Escarpment Showing Farm Holdings and House Holdings

Table 2 presents the typical extreme oscillation of runoff with high flood peaks after rainstorms in August, September and October when catchment conditions are saturated and rainfall intensity is high. The typical effects of torrents are fluvial erosion, slope erosion, landslides, earth and debris flow, discharge of large volumes of organic and inorganic solid urban waste into channels and flood plain zones, sedimentation and inundation. Both natural and anthropic factors contribute to the flood incidences. The principal natural factor is the rainfall intensity. (Table 3)

Table 2: Magnitude of catastrophic floods in selected streams

S/N	Flood Zone	Monitored site	Discharge (m^3/s)	
			Normal	Peak
1	Mugheb	Rota Bridge	3.0	1530
2	Sissia	Sessia Bridge	2.5	1020
3	Mulang	Mulang Bridge	4.2	1024
4	Old town (Sonac)	Sonac Bridge	2.2	352
5	Metta Quarter	Roundabout Bridge	1.3	1743
6	Musang	Dodretch Bridge	2.8	1489
7	Ntamulung Valley	Junction Bridge	0.9	470

Table 3 presents data on rainfall intensities during the months with flood incidences. Hawkins and Brunt (1965), using one year's data, estimated average intensities to be 101mm/hour. Even soils with rapid infiltration rates cannot possibly cope with such a downpour of water and large volumes of water are bound to gather in every depression. Where this is discharged into the urban landscape, soil, organic and inorganic wastes are swept away to congest drainage systems, natural waterways, channels and flood plain zones. This load exacerbates the flood problem. The average rainfall intensity of 168.5mm/hour is highly erosive and can generate enormous torrents and sediments.

Table 3: Estimated rainfall intensities in Bamenda

Date	Duration of fall of highest rainfall (minutes)	Rainfall amount (mm)	Rainfall Intensity mm/hour
4:8:2005	10	27.8	167
11:9:2005	15	41.2	165
14:9:2005	10	30.3	182
29:10:2005	10	26.7	160

The small upland watersheds in the urban field have since 1930 moved from moist, evergreen sub-montane forest to farmland and rough pastures. These areas and flood plain zones have been invaded by squatters with the following hydrological consequences of urbanization (Fig. 2):

√ clearing of forests for house and farm holdings in uplands and escarpment areas;

Fig. 2: Cross Section Of A Flood Plain In Ngomgham: **Urban waste receptacle, drainage and reclamation works, invasion by substandard householdings and often planted with sugar cane and *Xanthosoma* species, plantains and other fruit trees**

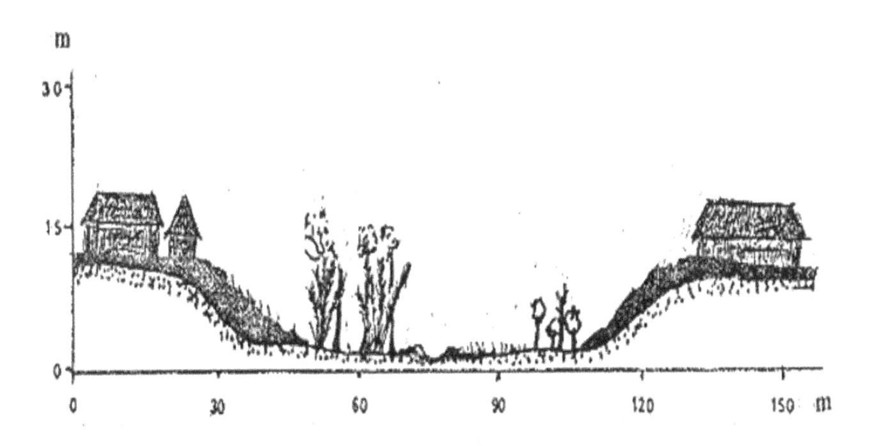

- √ relief clearance for civil engineering structures (earthworks and land forming);
- √ drainage and reclamation of flood-prone zones, swamps and wetlands;
- √ sinking of concrete foundations close to natural channels; concreting of pavements and pathways; construction of sub-standard houses in flood plain zones with the most insalubrious conditions and unaesthetic landscapes;
- √ change in drainage patterns: channels are straightened to speed up discharge; channel capacity is reduced by building close to stream banks; construction of bridges and channels unable to accommodate peak discharges; bank riveting; canals and makeshift dykes;
- √ congestion of flood plains, natural channels and drainage reticulation systems with solid waste; and
- √ farming in uplands, field drains and urbanization (drains and sewers) increase throughflow to streams.

Since the 1980s, the frequency and magnitude has been on the increase (Acho-Chi, 1998). According to Acho-Chi, the causes of increased flooding in urbanized watersheds are:

- increase in the percentage of impervious surfaces;
- paving, straightening or other ways of improving stream channels;
- landscaping (decrease in inundation area and surface removal of natural vegetation) and fragmentation of land into householdings; and
- filling-in and human occupation of flood plains.

Fig. 3: Cross-Section of Bamenda: Escarpment = Landslide and Torrent-Prone Zone. Lowlands = Riparian Flood, Sediment and Debris-Prone Zones

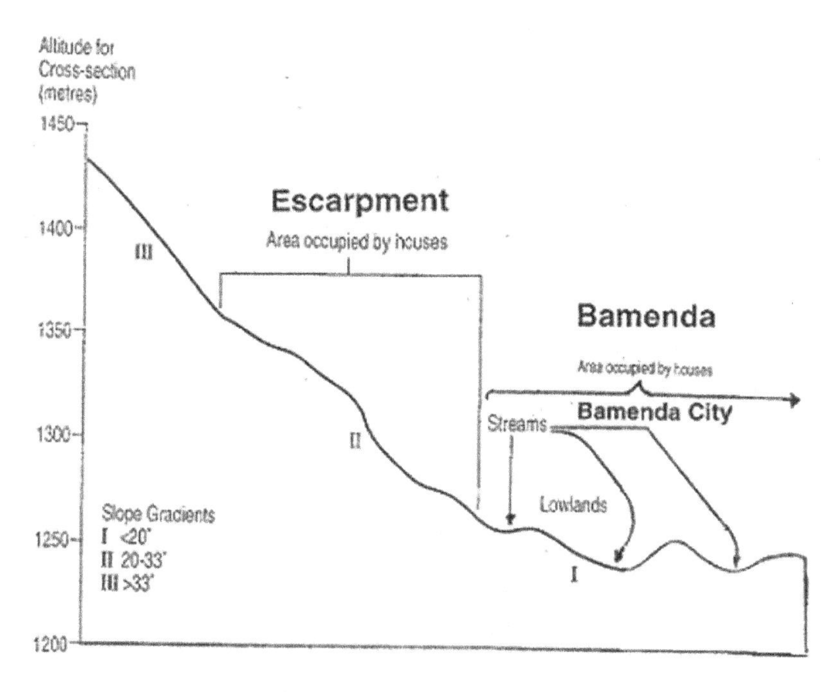

Intense rainfall in the upland riparian areas and escarpments often causes many woody debris flows and slope failures. Squatters have been killed by sediments. Tremendous crop residues, woody and soil debris yielded by slope collapse often hit houses. Consequently, woody debris, sediment, mud water, and urban solid waste inundate the flood plains in the valleys (Fig. 2 and Table 4). The slopes are composed of massive basalts which have relatively few joints and fractures. These are overlain by extrusions of trachytes which for the most part are strongly jointed. The general slope is between 20^0 and 33^0 with vertical upper slope segments. Below the vertical slope segments are rock outcrops, boulders, colluvial cones, deep ravines and gullies with urban squatters sprawling up the slope (Fig. 3).

Table 4: Sources and yield of soil and plant debris: landslides and debris flows (plate1)

S/N	Average length (m)	Average width (m)	Average depth (m)	Volume of soil mass (m3)	Year	Nature of debris: General observations
1.	30	8	12	2,880	2005	Woody debris and soil originating from farms in the escarpment zone and upland watersheds. Torrents and runoff sweep urban solid waste and carry it as debris. This congests bridges, culverts and drains, natural channels and flood plains, causing the buildup of flood water.
2.	80	20	30	48,000	200	
3.	60	9	10	5,400	1999	
4.	8	5	6	240	1998	
5.	12	10	15	1,800	1997	
6.	20	6	6	720	1995	
7.	50	22,5	30	337,500	1980	
8.	20	17	15	3,.060	1978	
9.	8	4	5	160	1978	
10.	20	6	6	1,200	1977	

Source: After Lambi (2004) and Ndenecho (2006)

Figure 4 shows a section of the escarpment with equal area stereograms of structural data along curvilinear slip surfaces observed after the August 1998 torrents and floods. The foothill zone has a basement of granite-gneiss. These are highly weathered with only pockets of fresh outcrops along cliff shelters. The upper slopes are highly fractured and traversed by joints that are quite outstanding at the slip movement. It is composed of remoulded clay with a consistency ranging from medium to stiff. The slides take place in lubricated joint planes that are suitably inclined to permit slope failure. There is no doubt that local intense rainfall combined with the geology triggers these disasters. Moreover, the rapid land development based on non-compliance with city planning are also trigger factors.

Fig. 4: Cross Section of Bamenda City Escarpment and Upland Watersheds Showing Equal Area Stereograms of Structural Data along Curviplanar Slip Surfaces: 1. Failure surface 2. Springs at the toe of slope in August-September- October 3. Highway 4. Creeping slope/gullies/farms. ., stream 6. Basement of granite-gneiss. (Source: Eze and Ndenecho, 2004)

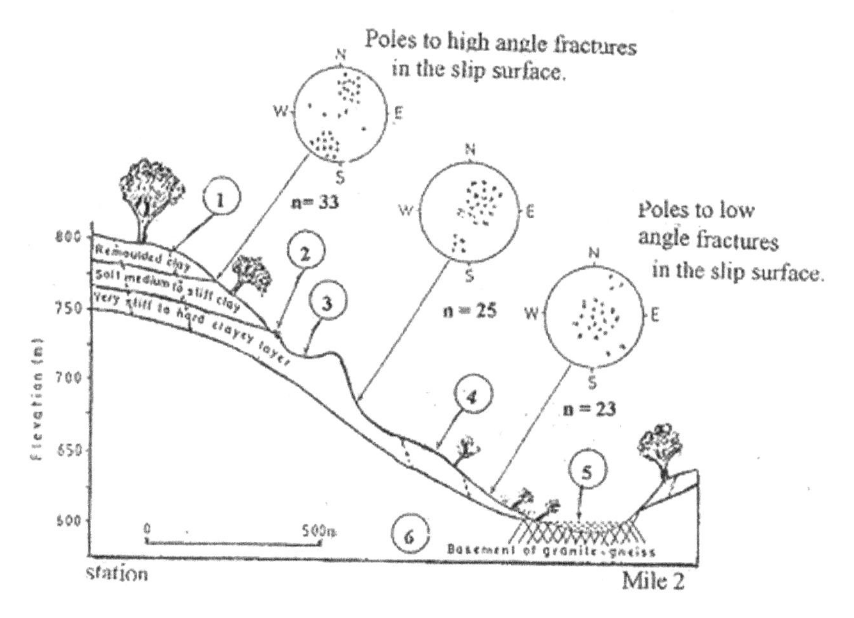

The sediments and debris are generally washed into flood plain zones by torrents and runoff. The streams deposit the greatest portion of their load close to the banks to form a natural levee. This wedge serves as a natural barrier against future floods. Figure 4 presents the cross-section of a flood plain below the Food Market. Squatters in flood-prone areas distort these levees by in-filling of valley walls through land drainage and reclamation. This, as shown in Figure 4, often results in the elimination of natural levees. It exacerbates the flood problems as flooding and aggradations recommence when these sites are levelled. Squatters have understood this. They now build higher levees with mounds of earth, stones, sandbags, used car tyres and makeshift materials. But these measures has always failed to prevent catastrophic events. Instead, they have backfired and intensified the severity of the flooding. Figure 5 demonstrates that water under high pressure within the constructed channels seeps beneath the levee embankment and bubbles up behind the levee walls on the flood plain, forming rising pools of water and swamps in July-August-September. This seeping water often undermines the levees in some locations, causing them to collapse at various points along the length. This discharges an outburst of water typical of the Mulang and Musang valleys.

Fig. 5: A Failed Levee in a Flood Plain Zone at Small – Mankon: Powerful floods tear away the levee wall on stream banks and together with seepage cause it to collapse: 2. Normal stream level. 2. Flooded stream. 3. Permeable sediment and solid urban waste. 4. Seeping water. 5. Levee built with solid urban waste, earth and sandbags (width=6mm, height = 1.5m). 6. Sloughing 7. Small drainage ditch. 8. Earth and sandbags, metal sheets and garbage form barriers. 9. Householdings on narrow flood plain (Source: Ndenecho, 2007).

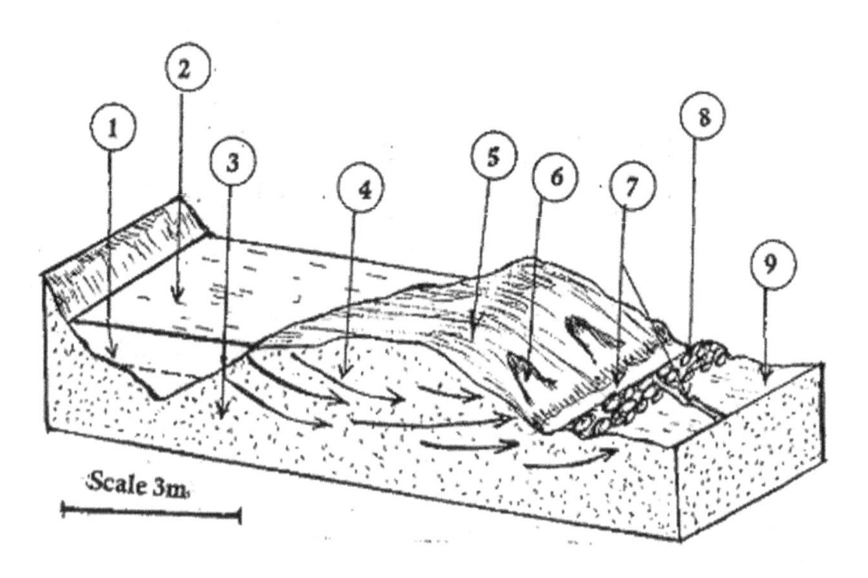

Conclusion

Floods accompanied by landslides and debris in flood plains and channels from August to October are triggered by high intensity rainstorm events. Urban poverty and the breach or absence of urban land use planning and zoning regulations have resulted in human occupation of flood and landslide-prone areas. A wide range of human actions over which humans have some control visit these disasters upon people every year, that is, these exacerbate flood and debris flow problems. Prompt land use regulations are required to restrict housing projects in the disaster-prone zones. Mitigation strategies have to be considered by the urban management authority. "Hard" engineering solutions will not be feasible for developing countries. These are expensive and the expertise may be wanting. Constant maintenance of structures is also a crucial problem. Poor countries are not in a good position to consider and adopt such a solution. There is, therefore, a need to search for alternative "soft" solutions. There is also a need to preserve the few flood plains left within the city. This will involve buying back additional wetlands from private owners and squatters and passing zoning regulations that discourage the development of hazard-prone areas. Some of these open spaces can be developed as parks, gardens and aesthetic ponds. Some stretches of flood plains can be zoned and allowed to flood naturally without posing any risk to life and property for scenic and ecotourism purposes. These must be accompanied by an effective solid waste management strategy. Such strategies can probably save human life, property and money in the long run.

Acknowledgements

Figure 4, originally published in Eze and Ndenecho (2004) is based on the field work of Mr. Titus Numfor (Geology Lecturer in the Cameroon College of Arts, Science and Technology, Bambili).

References

Acho-Chi, C. (1998) Human interference and environmental instability: addressing the environmental consequences of rapid urban growth in Bamenda. *Environment and Urbanization*, Vol. 10, No. 2, IIED, London, pp 161-174.

Chefor, I. (2000) Urban flood trigger mechanisms and flood consequences: Case study of Bamenda, Unpublished Long Essay, Department of Geography, University of Yaounde I, 27pp.

Dolgoff, A. (1996) Physical geology. D.C Health and Company, Lexington, pp 375-378.

Eze, B. and Ndenecho, E. (2004) Geomorphic and anthropogenic factors influencing landslides in the Bamenda Highlands, North West Province, Cameroon. *Journal of Applied Social Sciences*, Vol. 4, No. 1, pp 1526.

Hawkins, P. and Brunt, M. (1965) Soils and Ecology of West Cameroon, FAO, Rome, 516 pp.

Lambi, C. (1989) The problem of floods in urban Yoaunde. *Annals of the Faculty of Letters and Social Sciences*, Vol. 5, No. 2, University of Yoaunde, pp 175-190.

Lambi, C. and Fogwe, Z. (2001) Combating inundation in some major Cameroonian cities: an appraisal of indigenous strategies, in: C. M. Lambi (ed.) *Environmental issues: Problems and prospects.* Unique Printers, Bamenda, pp 133-154.

Lambi, C. (2004) A revisit of recurrent landslides on the Bamenda escarpment. Journal of Applied Social Sciences, Vol. 4, No. 1, University of Buea, pp 4-14.

Mainet, G. (1978) *Les innondations intérieures à la ville de Douala, Yaounde,* Department of Geography, University of Yaounde, 67pp.

Malcom, N. (1983) Hydrology; Measurement and Application. Macmillan, London,pp 39-42.

Ndenecho, E. (2006) Implications of slums on torrent processes: developing an environmental management plan for Bamenda city. *Journal of Environmental Sciences,* Vol. 10, No. 1, University of Jos, pp 81-88

Ndenecho, E. (2007) Landslide and torrent-channel problems of mountain slopes: processes and management options for Bamenda Highlands. Unique Printers, Bamenda, pp 46-58.

Ndoh-Nwi, T. (1996) Settlement on Bamenda escarpment: a contribution to urban geomorphology, Unpublished Long Essay, Department of Geography, University of Buea, 36 pp.

Tadonki, G.(1999) *Douala: Les exclus des marécages. Editions Mandara,* Yaounde, 117pp.

T'Hart, T. and Langeveld, J. (1995) The environmental situation in Bamenda, Cameroon. Institute for Forestry and Nature Research, Wageningen, 68 pp.

Tyler, M. (1985) Living in the environment. Wadsworth Publishing Company, Belmont, pp 218-220

Wijkman, A. and Timberlake, L. (1984) Natural disasters. : Acts of God or Acts of Man?, Earthscan, London, IIED, pp 64-69

Yasuhiro, D. (2001) Lessons from recent woody debris disasters in Japan, in: C. M. Lambi and E. B. Eze (eds.) *Readings in Geography*, Unique Printers, Bamenda, pp 107-118.

Chapter Four

Threats to the Ecological Stability of the Compound Farms in The Bamileke Plateau, Cameroon

Summary

In its agrarian civilization, humankind developed site-specific farming systems that were at equilibrium with local culture, socio-economic circumstances, and ecology. Production was oriented mainly to farm family subsistence. Due to rapid demographic growth, access to markets and the influence of foreign values these once sustainable agroecosystems are foundering. The paper investigates the structure, composition, and interactions of the compound farm system or home garden with local culture and ecology using a combination of field observations, ethnobotanical surveys, informal interviews and secondary data. It identifies the strengths and weaknesses of the system, the main threats to its sustainability and concludes that the compound farm is a gene bank for potentially useful but endangered plant species. The concept of the compound farm system as multipurpose gene banks requiring little inputs but enjoying protection by farmers who derive benefits from the system, may be the only way to minimize the current genetic erosion occurring in tropical montane forest ecoregions. The paper therefore recommends that agroforestry research programmes must emphasize the production of seedlings of these multipurpose species in the short-term. In the long-term, exploitation of variations in phenology both within and between species could result in longer term or all year round availability of desired vegetables, fruits, and other useful organs. There is need to enhance the productivity and sustainability of the compound farm system as a possible niche for the preservation of useful plant species.

Introduction

A closer look at the situation of tropical agriculture reveals that change has taken some paths. Originally, agriculture depended on local natural resources, knowledge, skills and institutions. Diverse site-specific farming systems evolved out of a long process of trial and error in which balances were found between the human society and its resources base. In most cases, production was oriented mainly to the subsistence of the family and the community (Reijntjes, *et al*, 1995). Traditional farming systems continued to develop in a constant interaction with local culture and ecology. As conditions for farming changed, for example, because of demographic growth or influence of foreign values, the farming system was also changed (Kapelle *et al*, 2000). Many traditional farming systems were sustainable for centuries (TAC/CGIAR, 1988). Mountain cultures and ecosystems face three primary threats: Land scarcity fuelled by inequitable ownership patterns and control of public resources, intensive resource extraction, and mass tourism and recreation.

However, these systems have had to cope with rapid changes during and since the colonial period: the introduction of foreign education and technology in agriculture and health care; increased population pressure; changes in social and political relations; and incorporation into an externally controlled international market system. The compound farms of the tropics need adjustments to these changes (Nair *et al*, 1986). The paper investigates the structure, composition and functions of the compound farm as a gene bank for potentially useful but endangered species, and identifies the scope for tapping and keeping alive this valuable indigenous knowledge about genetic resource management.

The Study Area And Problem Background

The Bamilike plateau is located in the West Cameroon Highlands which are an ample rainfall area in the Sudano-Guinean Savanna ecological zone. The highlands are horst – like mountains mostly with an altitude of 1500 metres above sea level. This is a region of plateaus of varying heights with lava accumulations forming the main mountains. Two main relief features are typical:

- There are basalt plateaus formed by the flow of basalt in Dschang (Foukoue), Mbounda, Bandjoun and Bazou. The study focuses on this area.
- Mount Bambouto which is a mountain block at the centre of these plateaux is greatly dissected by erosion and has an elevation of 2740m. There are many trachytic outcrops near the summit. The succession of volcanic stages seems to be ignimbrite-basalt-trachyte and phonolite (Tchoua, 1979).

The soils are Ultosols derived from basalts and trachytes with varying degrees of weathering. Precisely, the soils are acidic, low in major nutrients and have high phosphorus requirements. Due to population pressure, food crop fields are found on steep slopes where erosion losses are phenomenal leading to a decline in soil fertility. Using purchased inputs to overcome the above degradation scenarios in the traditional farming setting appears remote because farmers lack adequate cash and good input delivery networks.

The study focused on some villages in Bafoussam, Bandjoun and Dschang. This part of the plateau consists mainly of rounded hills with low-lying valleys separating them. The hills have an altitudinal range of 50 to 150 metres above the swampy valleys. Each hill slope section comprises a slightly convex crest-slope, a markedly convex upper slope, and an almost straight lower slope which ends abruptly with the bottomland. Almost the entire hill slope is utilized as farmlands characterized by hedge rows enclosing farm plots (Fig. 1).

There is an intense rainfall season from mid-March to mid-November and drought season from mid-November to mid-March. Annual rainfall varies between 1500mm and 3000mm but averages 2400mm with peak rainfall occurring between mid-July and mid-September. Temperatures fluctuate greatly and average 23^0 C. Although the area receives abundant rainfall which exceed 1800mm per year and which is distributed over a period of more than nine months, it is characterized by almost tree-less grasslands, fallow plots and a typical landscape of cultivated plots enclosed by live hedges.

The characteristics of the actual vegetation landscape have formed under the great influence of intense human activity (Letouzey 1968). See figure 1 for the population density. Letouzey, (1979) classifies the vegetation landscape as a "sub-montane domestic landscape derived from moist montane evergreen forest". Tamura (1986) demonstrates that agriculture on the plateau had been widespread by 2000 years. B. P. This resulted in widespread deforestation inducing soil erosion which is recorded in truncated regolith profiles, and the presence of artifacts in the migratory layers of profiles. The above considerations lead to the conclusion that forest had been considerably degraded by human activities by 2000 years B.P. Montane – related farming systems have therefore

Fig. 1: Location of Study Area and Rural Population Density in Bamileke Plateau (After Champaud, 1971): Inhabitants/ Sq. km.

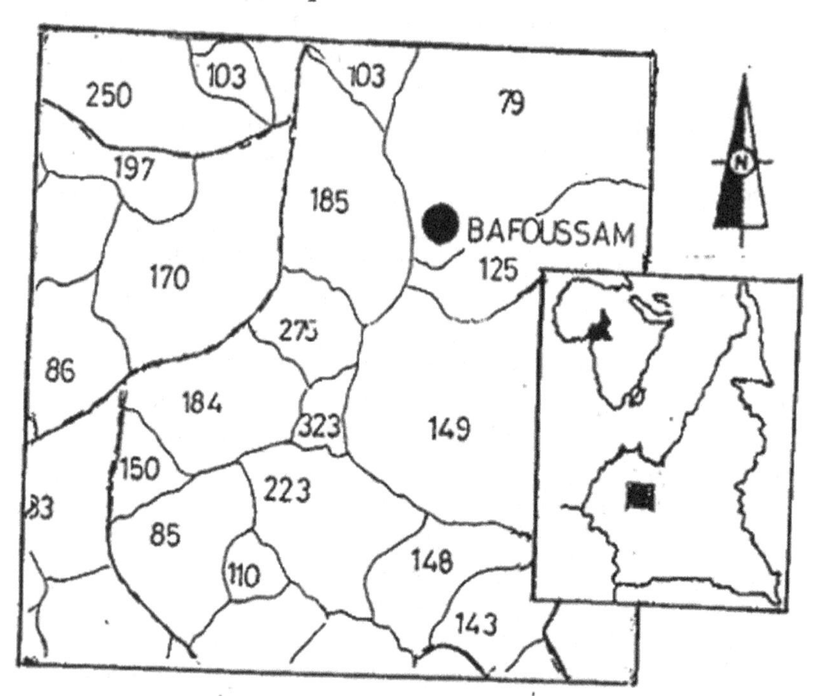

developed over the millennia as a result of interactions between high demographic pressure and the available natural resource base (Ngwa, 2001; Ndenecho, 2005) Compound farms are a traditional land use system that appears to have evolved from the shifting cultivation and bush fallow systems as a result of high population pressure and the need to establish individual land tenure. The system has been recognized as a potentially sustainable land use (Lagemann, 1977; Fernandes and Oktingali, 1984). It has possible application for the whole of the humid tropics. Figure 2 presents the possible nutrient cycles at the compound farm level. It presents the possible losses of nutrients, and natural gains of nutrients and management options. Within the farm level, nutrient flow is more or less cyclic. This guarantees the sustainability of the system because it seeks to optimize nutrient availability and cycling (Reijntjes *et al*; 1995). This once ecologically stable farming system which constitutes the last remnant of potentially valuable germplasm banks of traditionally important multi-purpose trees/shrubs of the montane and sub-montane tropical regions is threatened by rapidly changing socio-economic conditions resulting from land use pressure and increasing access to markets (Fernandes and Nair, 1986; Okafor, 1981, Okafor, 1982; Nair and Streedham, 1986).

Fig. 2: Model of Nutrient Cycle at the Compound Farm Level (After Reijntjes, et al, 1995, p. 66)

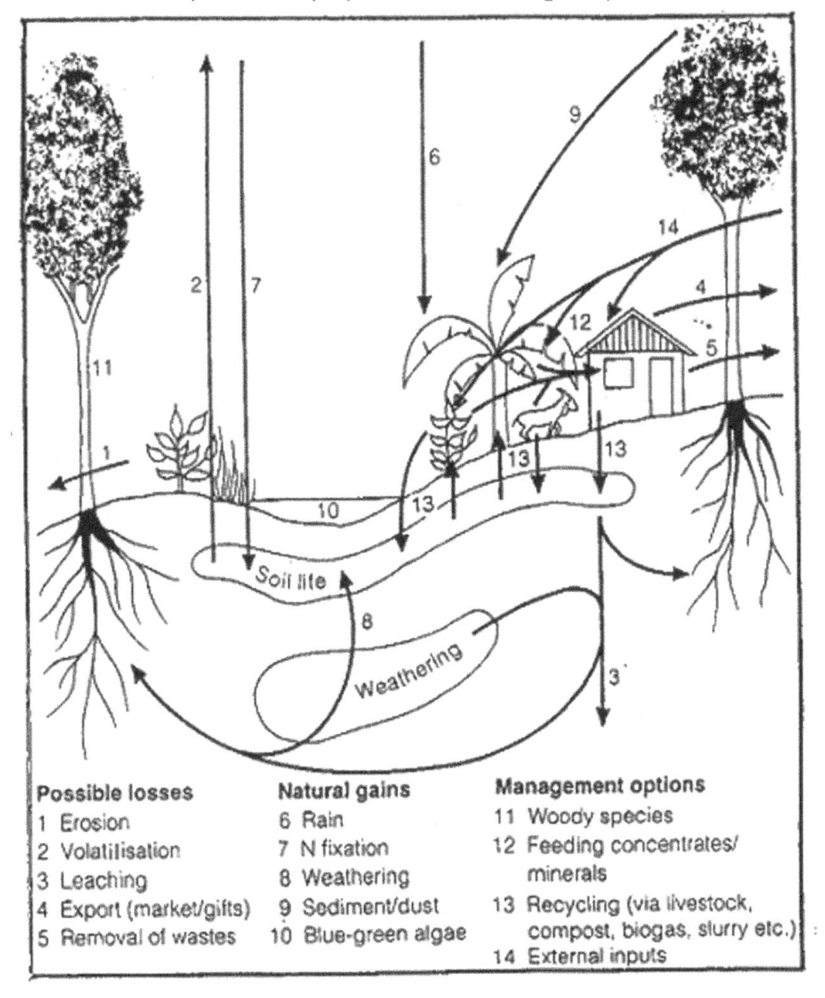

Methods and Data Sources

Based on the interpretation of aerial photographs covering Bamendjou, Penka-Michel, Bandjoun and the Koung-khi areas the structure and functioning of compound farms were investigated. The compound farm landscape was mapped in the Bamendjou area (Fig.3 and 4). A total of 244 compound farms were randomly selected in 7 villages for study. These were investigated in terms of the structure, functions and composition of compound farm components, management options and the major threats to the agroecosystem. A combination of field observations and informal interviews were employed. Floristic components of the system were identified and recorded together with the indigenous uses and functions indicated by each farm family head. The data obtained were complemented using secondary sources. The data so obtained assisted in the establishment of the ecological significance of the agroecosystem, the threats to the system and the scope for innovation.

Results and Discussions

Figure 3 presents a compound farm landscape in Bamendjou. Compound farms are found within the vicinity of homesteads and comprise numerous multipurpose woody species in intimate multistoried associations with annual crops and small livestock. The multistoried structure and species diversity allow almost complete coverage of the soil by plant canopies thereby promoting soil conservation (Beets, 1989; Asah, 1994; Okafor, 1981; Fernandes, 1984). The compound farm plots are enclosed by live hedges. Soil fertility is maintained by the use of household refuse, crop residues, animal manure, and thatching grass *(Hyparrhenia spp.)* from renewed house roofs after every 4 to 5 years.

It is a perennial cropping system involving rain-fed production. The average farm size is estimated at 1.2 hectares. Land tenure is by inheritance. The system is characterized by the perennial cropping of Arabica coffee, with over 95% of the plots intercropped with food crops. Bananas and plantains dominate, followed by maize, beans, cocoyams, colocassia, sweet potatoes and cassava. Root crops are markedly abundant. Fruit trees and other multipurpose trees are randomly scattered on the field and also are a main feature of the live fences. These provide shade and have other economic uses. The live fences constitute farm plot boundaries. The farm developmental stages are as follows: about 3 years to first crop, about 7 years to full bearing of coffee, which can continue up to 20 to 25 years. Trees are perennial and intercropping with annual food crops is typical. Replanting is done of individual bushes, parts of plots or whole plots when bushes die or show an important drop in yield (Tissandier, 1979). Crops are planted on mounds and ridges. Farmers bury and burn vegetation and crop residues to temporary restore soil fertility on food crop plots making sure that perennial crops and useful trees/shrubs are not killed.

Fig. 3: Live Fence Enclosures and Typical Hamlets in Bamendjou Area of the Bamileke Plateau (After National Geographic Institute Aerial Photograph)

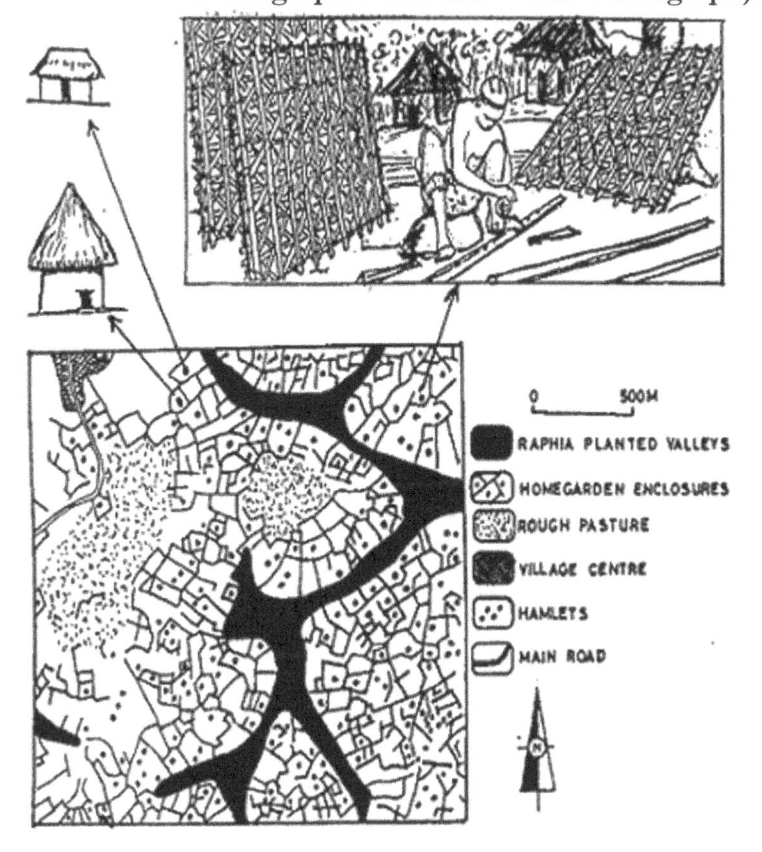

Table 1: Soil Fertilization Practices in the Bamileke Plateau

Crop Association	Percentage of farmers (%) out of a sample of 244							
	Organic Manure		Inorganic Fertilizers		Inorganic and Organic Manure		No Manure + No inorganic fertilizer	
	N	%	N	%	N	%	N	%
Coffee crop only	1	0.4	69	28.2	69	28.2	105	43.0
Mixed food crop /coffee	14	5.7	56	22.9	127	52.0	47	19.2
Food crops only	21	8.6	42	17.2	149	61.0	32	13.1
Subsistence crops only	51	20.9	10	4.0	27	11.0	156	63.9

The study investigated the soil fertility management practices in the system. Table 1 presents the adoption of practices by farmers. Inorganic fertilizer is used mainly for the coffee crop (28.2% of farmers) although it may also benefit food crops (17.2% of the farmers). The combination of inorganic and organic fertilizers (manure) is adopted by 61% of the farmers on food crops fields. On the whole, 43.0% of the farmers depend on natural soil regeneration for the coffee crop only, 19.2% for mixed food crop and coffee crop, 13% for food crops only and 63.9% of the farmers for subsistence crops only. Inorganic fertilizers are therefore used mainly for cash crops. It is therefore not longer a low external input agricultural system depending mainly on natural nutrient recycling through the association of useful trees and crops. Thatching grass for house roofing has rapidly been replaced by corrugated iron sheets. Compound farms cannot therefore be occasionally mulched with worn out thatching grass from roofs.

The land use system and structure of the compound farm is presented in Figure 4. The dominant staple crops include cocoyams *(Xanthosama sagitifolium), Colocassia esculenta, Dioscorea sp*, bananas *(Musa spp.)*, plantains *(Musa paradisiacal)*, and maize *(Zea mays)*. These are usually grown in mixtures with some subsidiary crops such as Okra *(Hibiscus esculentus)*, pumpkins *(Curcubita pepo)*, melon *(Cococynthis vulgaris)* and leafy vegetables. Sweet potatoes are an important cover crop. Cassava *(Manihot utilissima)* is a locally important crop. Fruit trees include mangoes, pears, pawpaw, oranges, guava, plums, *Canarium sp.* and raphia palms.

Several trees and shrubs are deliberately planted and managed on the compound farm for a variety of products or functions: These include the following:

- Compound farm products: food, timber, firewood, fodder, medicinal plants, manure, honey.
- Compound farm services: erosion control, soil conservation, shade provision, windbreaks, live fences against trespassers, crop protection.
- Soil amelioration: leaf litter serves as green manure, organic matter ameliorates soil structure and water infiltration rate. Nutrient recycling is facilitated by the multistoried association of plants.

Fig. 4: Compound Farm Landscape and the Structure of the Live Fence in Bamileke Plateau

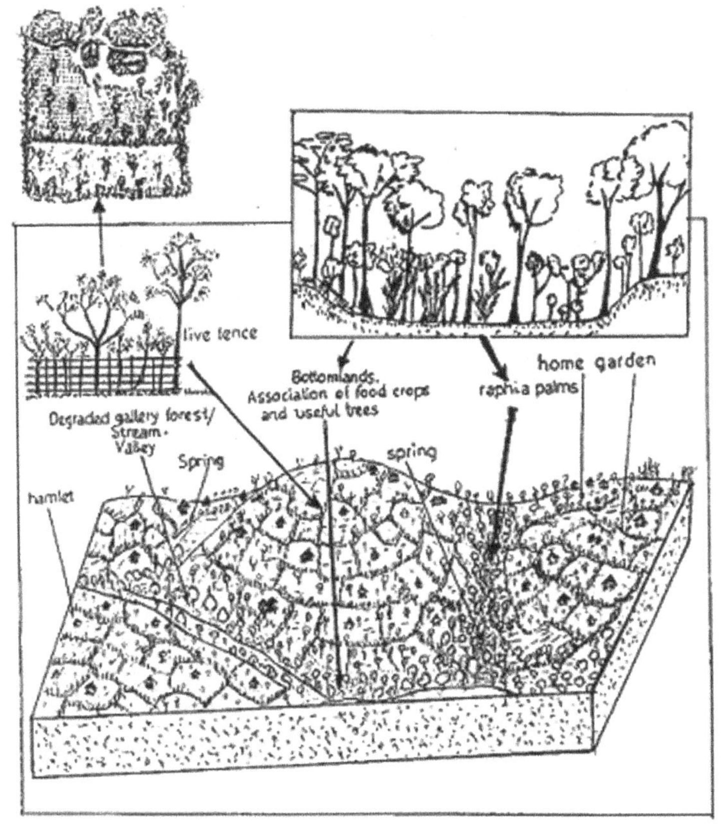

The landscape is characterized by a very intensive agriculture, with trees and shrubs fully integrated into the traditional arrangement of plants. The fences or hedge rows are composed of:

- Cuttings of fast growing trees and shrubs that are planted 0.8m to 1.5m apart. They serve as posts and later, some will grow to large trees;
- Other smaller plants to fill the spaces: 20 to 40cm apart; and
- An ordinary dead fence of raffia bamboo *(Raphia vinifera)*

The maintainance of the fence is done during periods when routine farm activities are less demanding in labour. It involves the following operations:

- Dead cuttings are removed and used as firewood and replaced by fresh ones at the beginning of the rainy season;
- Pruning of trees, except fruit trees and precious timber, higher than 1.5 to 2.0m, branches are used for fuel wood, leaves for green manure; and
- Routine maintenance and repair of the raffia bamboo fence.

The land use pattern varies with the soil catena. The following land use patterns can be identified: (Fig. 4):

- The hedgerows make a network of foot paths in between them. These paths allow sheep and goats to have access to natural pastures on the summit without encroaching on crop fields. However, with increasing demographic pressure, these rough pastures on marginal lands are now being converted to farm plots
- The homesteads are located on the middle slopes. The average farm is less than 1.2 ha. Less than 0.5 ha is planted only with food crops. About 0.7 ha are planted with food crops in association with coffee; and
- The valleys are planted with raffia palms and woodlots. These gallery swamp forest are now being drained for vegetable cultivation. The techniques used by farmers are different from conventional forestry. The farmers plant the trees in croplands and continue to cultivate around them until they form a closed woodlot. Many woodlots however appear to have been established to produce poles and timber rather than fire wood. However, woodlots have the

following functions:

- Products: food crops, timber, poles, fuel wood, fodder, litter, melliferous trees for bees, medicinal plants.
- Services: Soil stabilization, watershed protection and insect control, shelterbelts

Trees and Shrubs are essential perennial components of the compound farms. The potential of multipurpose trees and shrubs has been used by farmers to enhance sustainable crop-livestock production systems for centuries. Crop and rangeland form a mosaic of fallow plots, cultivated fields and open savanna with trees and shrubs as a regular feature. In these circumstances where there is a marked stratification of temperature with altitude, multipurpose tree and shrub germplasm is extremely varied and reflects current management. The main ecological niches in the compound farm systems where trees and shrubs are integrated are:

- Crop-based production system: croplands, fallow lands, compound farms, coffee plantations, compound hedges and boundary fences. The main vegetation types are: Legumes and shade trees, fruit trees, browses and live fences. Goats and sheep are main animal components.
- Mixed crop-livestock production system: farm lands, natural pastures, fallow lands, contour bunds, boundary fences, coffee plantations, compound farms. The main plant components are leguminous trees, and shrubs, fruit trees and browse plants. Goats, sheep and poultry are commonly kept for sale and for domestic consumption. Traditional farming practices include a combination of crop husbandry and small livestock. Each farm family keeps an average of 3 goats, 2 pigs and 5 fowls. Goats and sheep may range freely depending on the crop season. Fowls range freely while pigs are raised in pens. Livestock are fed with fodder from trees and shrubs, crop residues, grasses and herbaceous species growing in the compound farms and the woodlots. The woodlots are outlying fields. Eucalyptus woodlots are planted on hill tops while the bottomlands are planted with indigenous trees and possess the spring sources for homesteads (Fig. 4)

Table 2: Management of Compound Farm Live Fences in Selected Villages of Bandjoun Area

Village (code)	Enclosure with existing hedges	Enclosure with yearly mainte-nance	Enclosure with only casual mainte-nance	Enclosure with no mainte-nance	Hedges replaced with barbed-wire	Hedges replaced with exotic trees	Hedges replaced with cement blocks	Total
A	19	2	3	27	3	3	2	59
B	11	0	2	12	1	0	0	26
C	13	1	3	11	2	2	0	32
D	11	2	7	9	3	0	4	36
E	12	3	6	9	1	2	0	33
F	15	6	9	5	2	0	3	41
G	8	3	4	3	0	1	0	19
Total	89	17	34	76	12	8	8	244
Percentage	36.47	6.96	13.93	31.14	4.91	3.27	3.27	100

Table 2 presents data on the management of hedge enclosures of the compound farm. The table shows that 31.1% of live fences are not maintained, 13.9% receive only casual maintenance, 4.9% have been replaced with barbed-wire, 3.2% replaced with exotic trees (eucalyptus and cypress) and 3.2% replaced with cement block fences. Only 6.9% receive yearly maintenance. The compound farms of the Bamileke plateau are under increasing pressures. Although the present system has remained relatively stable over the years, present population pressure and scarce land are putting increasing pressure on it. Only 36.4% of compound farms are totally enclosed by live fences. With increasing monetarization of the traditional economy and the necessity to meet farm family basic needs, the emphasis of land use is turning to current productivity as opposed to sustainable productivity (Faha, 1999). The main Woody species used in live fencing which are now threatened are presented in table 3. Most of these indigenous trees are refuged in the gallery forests of the highlands. They are planted in live fences through cuttings. These plants species are rapidly being replaced in woodlots and live fences by exotic species. The gallery swamp forest refugia dominated by *Raphia* species are also rapidly being degraded (Dongmo, 1986).

Within the compound farm the spatial arrangement of annual and tree crops is typically random. Random mixtures have no specific pattern and often being the "arrangement" associated with shifting cultivation when trees are the result of random natural regeneration (seedlings and sprouts) rather than of systematic planting. The random arrangement is generally enclosed by border plantings or live fences (Table 3). The live fence encloses random stands of shrubs, biennials and crops. Border planting of trees is practised when short-statured crops cannot stand shading and where these crops are often of prime importance. The trees serve as boundary markers, as live fences and as fire and wind breaks; in addition the trees have other uses.

Table 3: Woody species recorded in live fences and hedgerows of the Bamileke plateau

No	Woody species in live fences	Other functions / uses
1	*Caesalpina Sp.*	Soil conservation, watershed protection.
2	*Cassia siamea*	Fodder, fuelwood, agroforestry.
3	*Cordia sp.*	Fodder, medicinal, carving wood, fuelwood.
4	*Croton macrostachyus*	Shade, timber, watershed protection.
5	*Crassocephylum manni*	Fodder, agroforestry, fuelwood.
6	*Dracaena sp.*	Traditional / ritual uses, boundary demarcation, erosion control.
7	*Daniella oliveri*	Agroforestry, fuelwood.
8	*Datura candida*	Fuelwood.
9	*Entanda abyssinica*	Fodder, fuelwood, erosion control, agroforestry, watershed protection.
10	*Erythrina sigmoides*	Medicinal, fodder, N-fixation.
11	*Embelia schimperi*	Fuelwood, Medicinal.
12	*Ficus Spp.*	Fuelwood, erosion control, boundary demarcation, ritual / traditional uses, fuelwood.
13	*Gmelina arborea*	Shade.
14	*Leucaena leucocephala*	Fodder, medicinal, fuelwood, N-fixation, reduces soil acidity.
15	*Maesopsis eminii*	Shade, timber, traditional / ritual uses.
16	*Markhamia tomentosa*	Erosion control, medicinal, timber.
17	*Schefflera barteri*	Medicinal, meliferous plant.
18.	*Schezolobium sp.*	Erosion control, fuelwood.
19	*Terminalia sp.*	Erosion control, fuelwood, shade, timber
20	*Vernonia sp.*	Edible leaves, erosion control.

The compound farm or home garden contains permanent combinations of annual crops, perennial food crops and forest trees with crown levels (multi-storey) – similar to a natural forest (Table 4). The striking feature of the compound farm is the continous economic and environmental role of the diversity of trees as they provide food for people, animal fodder, fuelwood, building materials, and "leaf manure" ; and at the same time, they stabilize and revitalize the soil. Farmers in the plateau have used this method for hundreds of years (Tamura, 1986) without any detrimental effects on the productivity of the land. The diversity of species accounts for stability of the system.

Table 4: Woody species recorded in compound farms in the Western Highlands

No.	Species	Propagation	Coppicing	Observations / Uses
1	*Acacia mearnsi*	Seedling	Yes	Fire resistant, erosion control, agroforestry, fuelwood, for restoration of landslides scars
2	*Albizia adianthifolia*	Seedling + direct seedling	Yes	Fodder, agroforestry, erosion control, fuelwood, fire resistant, watershed protection.
3	*Albizia gummifera*	Seedling + direct seedling	Yes	Fodder, agroforestry, erosion control, fuelwood, fire resistant, watershed protection.
4	*Albizia zygia*	Seedling + direct seedling	Yes	Fodder, agroforestry, erosion control, fuelwood, fire resistant, watershed protection.
5	*Caesalpina*	Seedling + direct seedling	Yes	Fodder, agroforestry, erosion control, live fence, watershed protection.
6	*Canarium schwei*	Seedling	No	Local fruit, shell, timber, not good on steep slopes, watershed protection, resin.
7	*Cassia siemea*	Seedling	Yes	Fuelwood, poles, melliferous, agroforestry

Table 4: Woody species recorded in compound farms in the Western Highlands (Contd)

No.	Species	Propagation	Coppicing	Observations / Uses
8	*Cassia spectabilis*	Seedling	Yes	Fodder, agroforestry, erosion control, watershed protection.
9	*Cordenia torrufolia*	Seedling	?	Fodder, watershed protection.
10	*Cordia milleny*	Seedling + direct seedling	Yes	Fodder, watershed protection, poles, timber
11	*Cola acuminate*	Seedling	No	Nuts, enrichment planting in watersheds and fields
12	*Croton macrostachyus*	Seedling	Yes	Live fence, shade, timber, poles, fodder, agroforestry, erosion control, watershed protection
13	*Cordia africanus*	Seedling + direct seedling	Yes	Fodder, live fence, erosion control, fuelwood, watershed protection, timber
14	*Cmellina sp.*	Seedling	?	Live fence, erosion control, fuelwood, timber
15	*Crassocephylum mannii*	Seedling	?	Fodder, agroforestry, live fence, fuelwood.
16	*Dracaena sp.*	Cutting/direct seedling	Yes	Live fence, stabilization of gully banks and landslides scar restoration, fire resistant

Table 4: Woody species recorded in compound farms in the Western Highlands (Contd)

No.	Species	Propagation	Coppicing	Observations / Uses
17	*Daniellia oliveri*	Seedling	?	Fodder, agroforestry, erosion control, watershed protection, fuelwood, timber, fire resistant
18	*Etanda Africana*	Seedling	Yes	Fodder, agroforestry, live fence, erosion control, fuelwood, fire resistant, slow growth.
19	*Etanda abyssinica*	Seedling	Yes	Fodder, agroforestry, live fence, erosion control, fuelwood, fire resistant, slow growth.
20	*Erythrina sigmoides*	Cutting + seedling	Yes	Live fence, agroforestry, fodder, erosion control, watershed protection, fire resistant, contour wattling
21	*Ficus ovata*	Cutting + seedling	Yes	Fodder, live fence, agroforestry, fodder, erosion control, watershed protection, fire resistant, contour wattling
22	*Ficus glumsa*	Cutting + seedlings	Yes	Fodder, live fence, agroforestry, fodder, erosion control, watershed protection, fire resistant, contour wattling

Table 4: Woody species recorded in compound farms in the Western Highlands (Contd)

No.	Species	Propagation	Coppicing	Observations / Uses
23	*Ficus vogelianum*	Cutting + seedlings	Yes	Fodder, live fence, agroforestry, fodder, erosion control, watershed protection, fire resistant, contour wattling
24	*Leucaena L.*	Seedling + direct seedling	Yes	Agroforestry, erosion control, fuelwood, watershed protection
25	*Gmelina arborea*	Seedling	?	Live fence, starter tree (pioneer). Good for landslides scar restoration
26	*Grevillea robusta*	Seedling	No	Melliferous, shade, rapid growth
27	*Khaya grandis*	Seedling	Yes	Reforestation of slopes, shade, timber, landslides scar restoration
28	*Lasiosiphonglaucus*	Seedling	Yes	Reforestation, shade, timber, watershed protection
29	*Maesa lanceolata*	Seedling	?	Agroforestry, erosion control, timber
30	*Maesopsis emini*	Seedling	Yes	Shade, timber, watershed protection
31	*Markhamia tomentosa*	Seedling	Yes	Live fence, erosion control, timber, watershed protection
32	*Mimosa scabrella*	Seedling	Yes	Pioneer tree. Good for landslides scar restoration, watershed protection.

Table 4: Woody species recorded in compound farms in the Western Highlands (Contd)

No.	Species	Propagation	Coppicing	Observations / Uses
33	*Newtonia buchananii*	Seedling	Yes	Rapid growth, reforestation, good for landslide scar restoration
34	*Piliostigma thoningii*	Seedling	?	Fodder, erosion control, fuelwood, timber
35	*Podocapus milanjianus*	Seedling	No	Reforestation of unstable moderate slopes, timber, watershed protection
36	*Polyscias fulva*	Seedling	?	Reforestation, melliferous, timber, watershed protection
37	*Prunus africanus*	Seedling	Yes	Reforestation of unstable moderate slopes, watershed protection, timber
38	*Pseudospodia microcarpa*	Seedling	?	Fodder, erosion control / slope stabilization, fuelwood
39	*Raphia vinifera*	Seedling	No	Wine, furniture, landslide scar stabilization, watershed protection
40	*Schizotobium*	Seedling	?	Erosion control, agroforestry, fodder, live fence, fuelwood
41	*Sesbania macrantha*	Seedling	No	Fodder, agroforestry, erosion control, fuelwood

Table 4: Woody species recorded in compound farms in the Western Highlands (Contd)

No.	Species	Propagation	Coppicing	Observations / Uses
42	*Sorindeia sp.*	Seedling	Yes	Reforestation of potential creep slopes, shade, timber, melliferous, watershed protection
43	*Spathodea sp.*	Seedling	Yes	Reforestation of potential creep slopes, shade, timber, watershed protection
44	*Terminalia sp.*	Seedling	?	Reforestation of potential creep slopes, shade, timber, watershed protection, fire resistant
45	*Tephrosia vogeli*	Seeds	No	Fodder, agroforestry, erosion control, watershed protection, rapid growth, pioneer plant for landslides scar restoration
46	*Trema orientalis*	Seedlings/ cuttings	?	Reforestation of degraded slopes, shade, timber, watershed protection
47	*Vitex diversifolia*	Seedling	?	Agroforestry, fuelwood
48	*Vernonia sp.*	Cuttings/ seedlings	Yes	Pioneer plant, good for landslide scar restoration, rapid growth, fire resistant, edible leaves

Source: Asah, 1994; Zimmermann, 1996; Ndenecho, 2006

There are four general categories of products obtained from the compound farm:

- Tree products: firewood, poles, gums, medicinal products
- Perennial cash or subsistence crops: fruits, nuts fibres.
- Staple food crops: cocoyam, colocassia, maize, pumpkins, sweet potatoes, groundnuts, yams.
- Forage and mulch: leaves of trees and grasses under the trees, house refuse.

The age-old stability of the compound farm is today threatened by the following:

- Soil tillage practices have remained traditional for centuries. This is resulting in degradation and poor yields.
- It is perceived as a primitive form of subsistence land use. This view has resulted in little, if any, resources devoted to its study and improvement;
- There is rapid replacement of indigenous tree and shrub species with rapid growing exotic species such as eucalyptus and cypress. This has diverse effects on soil and water resources;
- The drainage and utilization of valleys (swamp forests) is eroding the native species of trees and shrubs – especially the raffia palms (*Raphia sp.*). A consequence is that many spring sources are drying up; and
- Urbanization and access to markets have induced the conversion of compound farms to market gardens and monoculture of cash crops dependent on high external inputs.

On the other hand, the compound farm is a viable and sustainable farming system appropriate for the humid tropics. Due to demographic pressure, it is subject to stresses despite the merits of the system:

- Naturally, a variety of foods are available to the farm family throughout the year. It minimizes the risk of crop failure;
- It reduces the problems of food preservation and storage common in the humid tropics;
- Farm labour is evenly distributed throughout the year, although there may be some idle months;

- The multistoried system maximizes incipient sunlight necessary for photosynthesis. The majority of crops are shade tolerant; and
- It optimizes nutrient availability and cycling.

Conclusion

Due to the high rate of forest clearance in the area a large number of potentially valuable species are currently being lost. Current agroforestry research programmes must emphasize the production of seedlings of these multipurpose species in the short term. In the long term, exploitation of variations in phenology both within and between species could result in longer term or all year round availability of desired vegetables, fruits or useful plant organs. Similarly, intensive selection methods could result in increased yields and quality of products. There is also need for the screening of multipurpose trees/shrub species for valleys. A well coordinated and systematic research programme is required to obtain information relevant to enhancing the productivity and sustainability of the compound farm system. Local government agencies and non-governmental organizations are already collecting species of interest to soil fertility maintenance. There is need for scientists, government agencies and non-governmental organizations to collaborate with farmers in trying to improve the selection, conservation, and distribution of genetic resources so that valuable indigenous knowledge about genetic resource management can be tapped and kept alive. The concept of the compound farm system as multipurpose gene banks requiring little inputs but enjoying protection by farmers who derive benefits from the system, may be the only way to minimize the current genetic erosion occurring in tropical montane forest ecoregions.

Acknowledgements

The author acknowledges the French-English interpretation of qualitative data by Wafo Jean-Claude and Bogné André of the Regional College of Agriculture Bambili and the use of aerial photographs produced by the National Geographic Institute of Cameroon.

References

Asah, H.A. (1994) Potential of multipurpose trees and shrubs in traditional crop-livestock production systems of the Bamenda Highlands of Cameroon. In: *Proceedings of Agroforestry Harmonization Workshop, 4th – 7th April, RCA Bambili, GTZ / USAID / Helvetas Cameroon*, pp. 8-9

Beets, W. C. (1989) The potential role of agroforestry in ACP countries. International Centre for Agricultural and Rural Cooperation. Wageningen. pp.11-29

Champaud, J. (1971) Regional atlas – West 2, United Republic of Cameroon. ORSTOM Yaounde, Annex sheet 1-5.

Dongmo, J-L (1986) Amènagement et mise en valeur des grandes bas-fonds aux sols hydromorphes en pays Bamileke. In Kadomura, H. (ed.) *Geomorphology and environmental changes in tropical Africa: Case studies in Cameroon and Kenya. Hokkaido University.* pp. 95-106.

Faha, K. (1999) Dynamique des paysages dans une zone densement peuplée: cas de Bandjoun. Unublished postgraduate diploma thesis, Dept. of Geography, University of Yaounde I.

Fernandes, E. M.; Oktingali A. and Maghembe, J. (1984) The chega homegardens: a multistoried agroforestry cropping system on Mount Kilimanjaro, Northern Tanzania. *Agroforestry systems 2*, pp. 73-86.

Fernandes, E. M and Niar, P.R. (1986) An evaluation of the structure and function of tropical home gardens. *Agroforestry systems 6*, p.17

Kapelle, M.; Avertin, C.; Juarez, E.; and Zamura, N. (2000) Useful plants within a Camesino community in Costa Rican mountain cloud forest. *Mountain Research and Development,* Vol.20, No2, pp. 162-171.

Lagemann, J. (1977) Traditional farming systems in eastern Nigeria. Weltforum-verlag Munchen.

Letouzey, R. (1968) Etude phytogeographique du Cameroun. Paul Lechavalier, Paris, 511p.

Letouzey, R. (1979) Vegetation. In: *Atlas of the United Republic of Cameroon.* Jeune Afrique pp. 20-24.

Nair, M.A. and Sreedharan, C. (1986) Agroforestry farming systems in the homesteads of Kerela, Southern India *Agroforestry Systems 6.*

Ndenecho, E. (2005) Savannization of tropical montane cloud forest in the Bamenda Highlands. *Journal of the Cameroon Academy of Sciences, Vol.5, No.1* pp. 3-10.

Ndenecho, E. (2006) Ethnobotanical survey of Oku Mountain cloud forest, Cameroon. *Journal of Environmental Sciences vol. 10. No.2*

Ngwa, E.N. (2001) Elements of geographic space dynamics in Cameroon: Some analysis. ME Printers, Yaounde. pp. 10-13.

Okafor, J.C. (1981) Woody plants of nutritional importance in traditional farming systems of the Nigerian humid tropics. Ph.D thesis, University of Ibadan. 383p.

Okafor, J.C. (1982) Promising trees for agroforestry in Southern Nigeria. McDonald L.H. (ed.) *Agroforestry in the African humid tropics. Proceedings of a Workshop held in Ibadan:* 27th April to 1st May. 1981 UNU.

Reijntjes, C.; Haverkort, B. and Ann, W-B. (1995) An introduction to low-external input and sustainable agriculture. ILEIS, Leusden. pp. 6-66.

TAC/CGIAR. (1988): Sustainable agricultural production: implications for international agricultural research. Rome: FAO

Tamura, T. (1986) Regolith Stratigraphic study of late Quaternary environmental history in the West Cameroon Highlands and Adamaoua. In: Kodomura, H. (ed.) *Geomorphology and environmental changes in tropical Africa: case studies in Cameroon and Kenya. Hokkaido University.* pp. 45-59.

Tchoua, F. (1979) The land. In: *Encyclopedia of the United Republic of Cameroon.* Les Nouvelles Editions Africaines, pp. 79.

Tissandier, J. (1979) Land use: the Bamileke lands (Western Highlands around Bamendjou. In: *Atlas of the United Republic of Cameroon.* Editions Jeune Afrique. Paris, pp. 51.

Zimmermann, T. (1996) Watershed resource management in the Western Highlands. *Manual for Technicians.* Helvetas Bamenda

Chapter Five

Threats to Biological Diversity Management in the Mount Cameroon Region

Summary

The problems facing the sustainable conservation and management of biodiversity in Sub-Saharan Africa have tended to be defined in ways that do not lead to acceptable solutions. The paper uses a combination of primary and secondary data sources to identify the problems mitigating against a sustainable biodiversity management in Sub-Saharan Africa. It posits that both the problems and the solutions are built on economic foundations that need to be clearly understood. The most costly and least effective management strategy is to rely on state power. Most of the forest with protection status exists only on paper. The paper concludes that failure results from the fact that the rights which are denied forest-adjacent villages are so basic to livelihoods that enforcement is ineffective and imposes considerable social costs. Ill-adapted strategies that undermine rural livelihoods are bound to fail. Identifying the complex problems mitigating against sustainable management, the paper argues for elaborate new models for wildlife management. It recommends a holistic wildlife management model which simultaneously addresses the pillars of sustainability (economic, productive, environmental, social and cultural) using the community forestry approach.

Introduction

Today's threat to species and ecosystems is the greatest recorded in history (McNeely *et al*; 1990). Virtually all of them are caused by human mismanagement of biological resources, often stimulated by misguided economic policies and faulty institutions that enable the exploiters to avoid paying the full costs of their exploitation. Solutions to the problem of biodiversity degradation depend above all on how the problem is defined. It appears that the problems facing the conservation of biological diversity in Sub-Saharan Africa have tended to be defined in ways that do not lead to acceptable solutions. The problems are generally defined in terms of insufficient areas, excess poaching (Jaff, 1994), poor law enforcement, land encroachment and illegal trade (Ndenecho, 2005; Denniston, 1995, Balgah, 2001).

These definitions warrant possible responses which include establishing more protected areas, improving standards of managing species and protected areas, enacting national forest protection laws, enacting international legislation controlling trade in endangered species and policing of protected areas. These measures are all necessary but they respond to only part of the problem. Fundamental problems lie beyond protected areas which affect the livelihoods of those who mismanage the natural resource base. This paper seeks to identify the problems mitigating against biodiversity management by appraising the current management strategies and to identify alternative strategies for Sub-Saharan Africa.

The Study Area

The study area is located between latitudes 4° N and 6^{0} 50‘ N and longitudes 8^{0} 50’ E and 10^{0} E. It covers a surface area of 24,910 Km^2 (Fig.1). Geographically, this ecoregion encompases the mountains and highland areas of the border between Nigeria and Cameroon excluding the Bamenda Highlands. It covers the Rumpi Hills, the Bakossi Mountains, Mount Nlonako, Mount Kupe and Mount Manenguba. (Stuart 1986; Gartland, 1989; Scatterfield *et al.*, 1998). Mount Cameroon is the dominant geographical feature with an altitude of 4100m. Most of the region is below 2000m in elevation. At about 800m to 1000m the ecoregion grades into lowland vegetation communities of other ecoregions. In the majority of cases, however, the lower boundary of these forests is now determined by conversion to agricultural land.

Rainfall is around 4,000mm per annum, declining inland to 1,800mm or less. The mean temperatures are below 20^{0}C due to the effects of altitude. This is a volcanic region. Soils derived from volcanoes are fertile, which makes the land attractive to farmers. Combined with adequate rainfall, this contributes to a high human population density. In White's (1983) phytogeographical classification, these mountain areas fall within the Afromontane ecoregion. It has several endemic vascular plants and reptiles. Nine of the reptiles are considered narrow endemics (Stuart 1986). In addition to the narrow endemics, there is also a significant overlap between the flora and fauna of the mountains of this ecoregion. There are 50 endemic plant species and 30 near endemic plant species (White, 1983).

Human activities are increasingly fragmenting, degrading and isolating the remaining forest patches. The Bakossi Mountains have at least 200 km^2 of mid-altitude and montane forest above the altitude of 1000m; and the lowland forest (Western Bakossi) covers some 400 km^2. The study concentrated on the Western Bakossi forest: Banyang – Mbo forest reserves (42.606 ha) created in 1936, Bambuko Forest Reserve (26,677 ha created in 1950, Southern Bakundo Forest Reserve (19,425 ha) created in 1940, Mokoko River forest Reserve (9,065 ha) created in 1952, Bakossi Forest Reserve (5,517 ha) created in 1956, Meme River Reserve (5,80 ha) created in 1951 and Barombi – Mbo River Reserve created in 1950 (Asong, 2001). This is one of the least well protected eco-regions in Africa where local traditional rulers still exert considerable authority over land use. The main section of Bakossi (550 sq. km) became a "protection forest", banning all logging in 2000. Kupe became a "strict nature reserve". The boundaries were delimited by the participation of local people in 2000 – 2001.

The main objective is to ensure sustainable resource management with emphasis on natural regeneration, the involvement of the local population and the promotion of sustainable livelihoods by implementing the 1995 wildlife management laws and guidelines (Fonyam, 2001). The law recognizes the creation of buffer zones around all protected areas. These are regions adjacent to protected areas, which provide local communities with sustainable income generating activities (hunting, selective logging, fuel wood and non-timber forest products) (Jongman, 1995; Kelkit *et al*; 2005):

- Zone 1: a natural zone where nature has priority of protection.
- Zone 2: a cultural zone sustaining local peoples livelihoods
- Zone 3: an intense usage zone with many land use systems.
- Zone 4: a rehabilitation zone with emphasis on restoration of biodiversity and natural landscapes.

Fig 1: Location of the Study Area and the Protected Areas of South Western Cameroon

Research Methods

The study used a combination of primary and secondary data sources to investigate the micro-incoherencies and constraints to biodiversity management. The conservation efforts of wildlife management workers were assessed by administering a questionnaire to 49 workers. This questionnaire focused on the main wildlife conservation activities and how they relate to the management strategy of the protected areas. Local communities adjacent to the protected areas with forest-dependent livelihoods were also investigated. Due to the large size of the study area, it was divided into five zones to ease data collection. These zones were:

- Zone A: villages in Mbonge area
- Zone B: villages in Munyenge in Muyuka
- Zone C: villages in Nguti and Bangem
- Zone D: villages in Bayang and Manyu
- Zone E: villages in Kupe Manenguba

240 questionnaires were administered in each zone to forest-adjacent residents. A total of 1200 questionnaires were administered. The questionnaire focused on the identification of livelihood activities, quantitative and qualitative assessment of livelihood activities , forest users per protected area and buffer zone and forest exploitation techniques. Field visits involved the use of Participatory Rural Appraisal, and semi-structured interviews to obtain information on user groups of the forest. These tools were also used to identify the human activities and impacts on the forest. Aerial photographs were used to complement qualitative data on land use impacts on the Kupe forest, and the South Bakundu Forest Reserve. The study was complemented by secondary data and field observations. Of the 1200 copies of questionnaires distributed to the local population, a total of 934 were returned (77.8%):

- 1140 questionnaires were distributed to the local population and 885 were returned (77.6%), and
- 60 questionnaires were distributed to wildlife conservation technicians and 49 were returned (81.7%)

Rural livelihoods dependent on the protected areas were investigated using a combination of field observations of the Mount Kupe and South Bakundu areas. Land use maps from these field sites were facilitated by the work of Effange (2006), Ewane (2006) and Mesumbe (2001). These were updated using a Global Positioning System (GPS). Data on livelihoods dependent on the protected areas were obtained from secondary sources (Ewane, 2006; Effange, 2006;) and the medicinal plant collection of the Limbe Botanic Garden which is yet to be published. The data so obtained was analyzed using descriptive statistical techniques such as tables, frequencies and percentages in order to establish the constraints and micro-incoherencies in biodiversity management in traditional societies.

Presentation of Results and Discussions

Table 1 presents the main activities of the forest conservation workers. Their main function is routine policing and guarding of the protected areas (57.1%). There is little research on species identification which engages only 18.4% of the workers. Routine boundary demarcation and regeneration/rehabilitation of degraded area suffer from acute shortage of workers. The 49 workers identified include 10 engineers, 18 forestry technicians and 21 forest guards. With such a weak staff base insufficient attention is given to the socio-economic and cultural situation of local people whose livelihoods depend on resources found in the protected areas. Attempts by development agents to protect or rehabilitate ecosystems in a particular place are often contradicted by the socio-economic pressures of local people. Many laws and regulations governing land use in protected areas remain unenforced.

Table 1: Main activities of wildlife conservation workers in Bamboko Forest Reserve and Bayang –Mbo wildlife sanctuary

Main Activities		Frequency	Percentage
1	Policing and guarding of protected area	28	57.1
2	Routine boundary demarcation/clearing	04	8.2
3	Research and Species identification	09	18.4
4	Regeneration/rehabilitation of degraded areas	06	12.2
5	No job specifications	02	4.1
Total		**49**	**100**

There is the lack of adequate consultation and clarification concerning the demarcation of boundaries of protected areas as well as the limited capacity of the workers to enforce the protected area status (Jaff, 1994). All these generate conflicts and it is now being realized that the "conservationists" approach to biodiversity conservation has failed to come to grip with crucial social issues. (Ndenecho, 2005b). This approach has provoked social conflicts which often undermine the possibility of implementing and achieving basic conservation objectives. Coupled with limited human and financial resources necessary for the administration of protected areas, most forests with a reserve status often exist only on paper (Kamanda, 1994). Under these circumstances, the non-enforcement of regulations becomes an explicit strategy for the state to reduce conflicts (Fig. 2 and 3).

Fig 2: Degradation Of The South Bakundu Forest Reserve by Adjacent Village Communities.

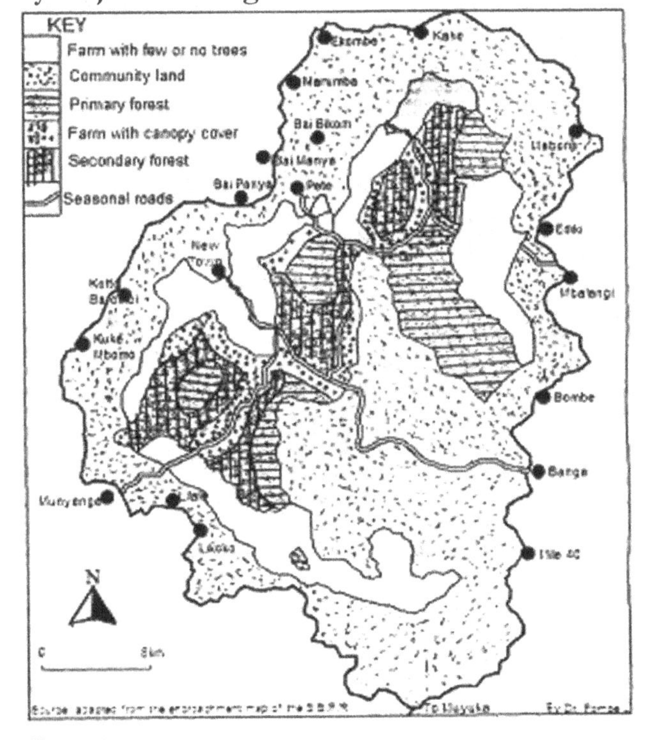

Source: Effange, 2006

Fig 3: Degradation of Mount Kupe Forest: A = primary forest, B = Secondary or disturbed forest, C1 = Farms with a less than 50% canopy cover, C2 = Farms with few or no trees,?= Areas yet to be mapped

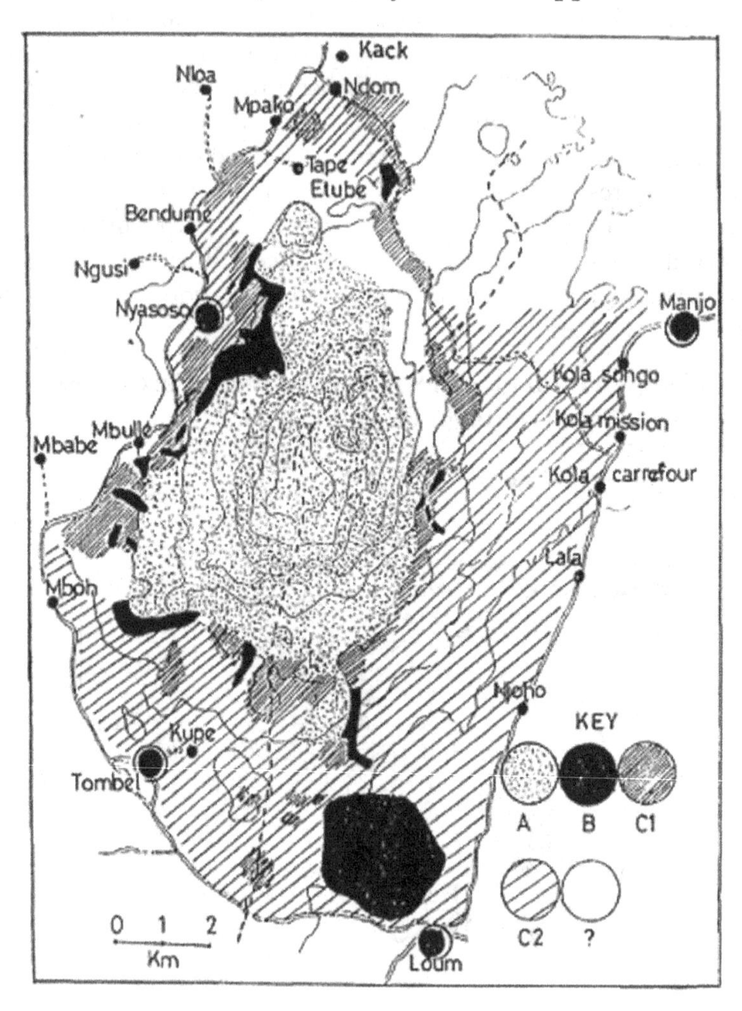

Source: Mesumbe, 2001

Table 2: Commercial non-timber forest Products (NTFPs) in protected areas

Name	Life form	Edible portion	Uses
Piper guineense	Climber	Seed	Flavour
Afromomum melgueta	Herb	Seed	Flavour
Afromomum logiscarpum	Herb	Seed	Flavour
Zingiber officinale	Herb	Rhizome	Flavour
Monodora myristica	Tree	Seed	Flavour
Ricino heudelotii	Tree	Seed	Flavour
Afrostynax lepidophylus	Tree	Bark	Flavour
Irvinga gabonensis	Tree	Fruit/seed	Juice/flavour
Poga oleosa	Tree	Seed	Kernel eaten
Gnetum africanum	Climber	Leaf	Vegetable
Allium cepa	Herb	Seed	Flavour
Piper nigrum	Climber	Seed	Flavour
Capsicum annum	Herb	Fruit	Flavour
Allium sativum	Herb	Bulb	Aroma
Occimum basillicum	Herb	Leaf	Aroma
Xylopia quintasii	Tree	Seed	flavour

Source: Effange(2006) and Ewane (2006)

Table 2 above presents the commercial non-timber forest products of the study area. These products are grown in the wild. These spices and vegetables are potentially important and are capable of meeting a wide range of tastes. They are therefore gaining acceptability both on the local and world market. They are also noted for their versatility in the preparation of many local dishes and medicines. These products grow in the wild in combination with other forest trees. In fact, most of the climbing types are found climbing on forest trees as support. They are rarely cultivated but may exist as protected stands on farmlands. Occasionally, some species are introduced as intercrops with arable or cash crops. They also occur in low densities in the wild and consequently the harvested quantities are low and the prices even at local markets are high. As these species are not generally planted, they are therefore very susceptible to genetic erosion. Forest – adjacent villagers must harvest these products for a livelihood.

Table 3 above presents the medicinal and economic values of plant species found in the area. These plants play an important role in highlighting the value of all plant to mankind. Most forest adjacent village communities in the area have been known as centres of traditional healing. A variety of plant parts are harvested for medicinal purposes and are commonly traded on the local and regional markets. *Nauclea diderrichii* is sold locally in sachets for the treatment of typhoid. The mercurial value of the timber is 8000 CFA/m^3. According to forestry department statistics in Cameroon, exploitation of *Pygeum africanus* in 1990/1991 was 1,121 tons, in 1989/1990 was 1,024 tons and in 1988/1989 was 726 tons, *Pygeum* bark made up 88.6% of the plant material brought to the local pharmaceutical factory (Plantecam) in Mutengene. Plantecam steam-dries bark 50% is exported in powder form post-drying and the other 50% is sold as extract. The Italian Company – Invemi Delta Beffa also markets pygeum products in Europe. According to Forestry Department statistics, 1989/1990 the exploitation of *Rauwolfia vomitoria* was 15 tons. Plantecam currently exports *R. vomitoria*. *Strophanthus gratus* compounds are used by the international pharmatceutical industry. This species is exported in relatively large quantities. In 1990/91, the quantity exported was 2.7 tons. Plantecam currently exports these species.

Plantecam extracts the seed oil of *Voacanga africana* which contains vincamin for export. In 1990, forest-adjacent villages delivered 900.6 tons of seed to Plantecam Mutengene factory. *Ancistrocladus Korupensis* was first collected in the Korup National Park by botanists under contract from the United States National Cancer Institute's Natural Products Branch. It was screened for effectiveness against cancer and found to respond even better to AIDS in preliminary trials. It has thus generated international pharmaceutical interest. It has since yielded the potential anti-HIV compound michellamine b. It is, however, still too early to know whether michellamine b will pass all the tests necessary to make it into a pharmaceutical drug (Effange, 2006).

Table 3: Medicinal Plants Found in Protected reas and their Economic Value

Plant species	Life form	Plant parts used	Observations
Afromomum spp.	Herb	Fruit, leaf, stem, Seed	Sold in local and regional markets
Alstonia boonei	Tree	Bark, latex, leave	Bark sold in localMarkets
Ancistrocladus korupensis	Climber	Leaves	Preliminary trials by scientists forthe cure of AIDSand Cancer
Annickia chlorantha	Tree	Bark, leaves	Sold on local Markets
Baillonella toxisperma	Tree	Bark, seed oil	Sold on localMarkets
Bryophyllum pinnatum	Herb	Leaves, fruit	Not sold in Markets
Canarium schweinfurthii	Tree	Fruits, seed, resin Bark	Fruits sold on local markets + high value timber 40.000CFA/m^3
Ceiba pentandra	Tree	Leaves, bark, root	Not sold in markets. Timber sold:8000CFA/m^3
Cola spp.	Tree	Seeds, leaves, Bark, roots	Sold locally and cola exported to Nigeria
Costus aferker	Herb	Stem, root, leaves Rhizomes	Not sold in Markets
Elaies guineensis	Tree	Sap, wood, leaves Oil and trunk apex	Oil, kennels, and Fruits sold
Eremomastax speciosa	Herb	Leaves	Generally not Sold in markets
Garcinia kola	Tree	Seed, root, bark, Latex	Seeds and barks Sold on local Markets

Table 3: Medicinal Plants Found in Protected reas and their Economic Value (Contd)

Plant species	Life form	Plant parts used	Observations
Garcinia mannii	Tree	Branches for chewing sticks, bark, leaves, latex	Bark sold. There is significant trade in chewing sticks
Kigelia africana	Tree	Buds, bark, fruits	Commonly sold in markets and regularly bought
Milicia excelsa	Tree	Exudates, bark, leaves, roots	Not sold, wood used for poles and furniture
Nauclea diderrichii	Tree	Bark, root, fruits	Branches sold for chewing sticks. Valuable timber
Newbouldia laevis	Tree	Bark, root, leaves	Sold in local Markets
Physostigma venenosum	Shrubby Climber	Seeds	Not sold
Piper guineensis	Woody Climber	Fruits, seeds, Leaves, roots	Commonly sold in local markets
Prunus africanus	Tree	Bark	Bark traded in World market
Pterocarpus soyauxii	Tree	Stem, bark, leaves	Leaves sold as vegetable, stem and bark sold
Rauwolfia vomitoria	Tree	Sap, seeds, leaves bark	Sold for industrial transformation
Ricinodendron heudelotii	Tree	Seeds, leaf, bark, root, kernel	Seeds sold widely in markets
Senna alata	Shrub	Leaves, bark	Not sold in markets
Spilanthes filicaulis	Creeping herb	Leaves, flowers	Not sold in local Markets

Table 3: Medicinal Plants Found in Protected reas and their Economic Value (Contd)

Plant species	Life form	Plant parts used	Observations
Strophanthus gratus	Shrub	Leaves, roots	Sold in the world market (exported)
Tetrapleura tetraptera	Tree	Fruit, seeds, bark	Fruit sold as a Spice. Bark not marketed
Voacanga africana	Tree	Seeds, latex, bark Root	Sold to industries. Exported

Source: Medicinal plants of the Limbe Botanic Garden (unpublished)

Since medicinal plants are found in the wild and mostly in concentrated stands in forest reserves, they are foraged by local people and sold to commercial companies. These commercial exploiters include Plantecam Mutengene Factory and International Transactions Trade Company in Douala. Each exploiter is assigned an area of exploitation and is prohibited from other areas. Permits are issued by the Forestry department, stating the area of exploitation and limiting the amount of bark to be extracted. If any company exceeds the limits imposed, then it has to pay re-aforestation taxes and a fine. Despite these conditions, this sector is characterized by the corruption of forestry officials by the companies and unsustainable harvesting techniques by harvesters (forest adjacent dwellers). Natural resource management systems through protected areas and buffer zones therefore generate issues of conflict with local communities.

Table 4 indicates how the local people derive their livelihood from the forests. Livelihood activities include hunting of game, fishing, farming, timber exploitation and the collection of non-timber forest products. For the two protected areas approximately 1040 people depend on the forests for their livelihood activity. These activities are very intensive in the protected area. Natural resources in the buffer zones have been degraded through unsustainable harvesting techniques. Livelihood activities are therefore encroaching into reserves and protected areas. Non-Timber Forest

Products (NTFPs) collection, fishing, hunting and timber exploitation rank highest. Most people keep totems and shrines in the buffer zone. These are well protected by indigenous beliefs and traditions.

Table 4: Population adjacent to Bamboko and Bayang-Mbo Forest Reserve with forest-dependent livelihoods

Livelihood activity	Total	Protected Area		Buffer Zone	
		Number	%	Number	%
Hunting	885	576	65.1	309	34.9
Fishing (river fishing)	785	560	63.3	225	25.4
Farming	889	553	62.5	332	37.5
Rituals	1040	538	43.3	502	56.7
Shrine/totem	885	347	39.2	538	60.8
Timber exploitation	885	500	56.5	385	43.5
NTFPs collection	885	770	87.0	115	13.0

Source: 2006 Fieldwork

Table 5 presents data on forest exploitation techniques per livelihood activity. Hunting involves 885 people and includes both subsistence and commercial hunting. Hunting has caused the decline in wildlife population in most protected areas (Jempa, 1995) and continued to pose a threat to the remaining populations of chimpanzee (*Pan troglodytes*), drill monkeys *(Mandrillus leucophaenus)*, forest buffalo *(Syncerus caffernamus)*, bush pig *(Potamochoenus porcus)*, giant pangolin *(Phataginuis fricupis)*, and the red-eared monkey *(Cercopithecus erythrotitis)*. Slash – and – burn cultivation involves 889 farmers. Sixty two percent of these farmers are encroaching on protected areas while 34% farm in the buffer zones.

Table 5: Forest Exploitation Techniques per livelihood activity in Bamboko Forest Reserve and Bayang-Mbo Wildlife Sanctuary

Livelihood activity	Implementation techniques	Respondents Involved	
		Number	%
Poaching	. guns	339	38.3
	. traps	328	59.7
	. others	18	2.0
Farming	. slash – and – burn	889	100.0
NTFPs collection	. gathering	799	90.3
	. cutting/ felling	18	2.0
	. ring – barking	68	7.7
Totems/shrines	.medicinal plants in shrine	594	69.1
	. animals as totom	291	32.9
Timber exploitation	. opening of forest roads	188	21.2
	. felling of trees	697	78.8
Fishing (River fishing)	. bamboo traps	226	25.5
	. fishing nets	69	7.8
	. hooks	590	66. 7

Source: 2006 field work

The increasing population and the need for cash is increasingly orienting the local economy from subsistence to commercial farming. NTFP collection involves unsustainable harvesting techniques. Trees are ring-barked, exposing the stems and leading to death. This constitutes a threat to the use of many medicinal plants as future resources. Tree deaths are partly due to illegal exploitation but many also are due to low wages which encourage company employees to extract as much bark as they can. The study identified 885 timber exploiters in the sites selected for study. Trees are felled and sawn using motorized saws. Timber is sold in urban markets. The opening of roads into the buffer zones and protected areas is a preliminary phase to the incursions of other forest degradation activities. The species harvested include: *Entandrophragma angolensis, Terminalia mollis, Pterocarpus soyauxii, Gibourtia tessmanni, Aninidium mannii, Entandrophragma cylondricum* and *Baillonella toxisperma.* The forest sector contribution to the gross domestic products and the weight of their exports is enormous about 4 to 6% per year (Effange, 2006).

Conclusion and Recommendation

Experience has shown that the most costly and least effective mechanism for enforcing rights, on the whole, is to rely on state power (Denniston, 1995; Stuart, 1986; Ndenecho, 2005). The difficulties of enforcing prohibitions on the use of state forests have been noted in many Sub-Saharan countries where most forests with protection status exist only on paper. Failure results from the fact that the rights which were denied forest-adjacent village communities were so basic to livelihoods that they were ineffective and imposed considerable social costs. Forest protection projects that ignore local livelihoods are bound to fail. This study recommends that community sanctions are the ideal mechanism for the "enforcement" through voluntary agreement of rights within fairly small and defined communities. (Tucker, 2000). They are more likely to put the local people's proprieties first, to be effective, and to be sustainable in economic and social terms. Arnold and Campbell (1985) summarized the many types of possible control systems which are used in traditional forest management by local communities. These are presented in table 6. The control systems are elaborated using participatory approaches and enforced using indigenous religion and other traditional institutional mechanisms.

The critical issue is not so much what rules are applied, but the strength of community institutions, which set the rules and ensure that they are effective. Community sanctions are most likely to arise spontaneously and work where a cohesive social administrative structure exists. Success is likely to be achieved where there are relatively isolated village communities, little affected by migration either to or from the area. Traditional religion and well-structured socio-political institutions in the village also offer mechanisms in favour of community forest. The land, water and biological resource base of a village community can be healed through holistic wildlife management. It involves the use of practical decision-making that effectively deals with complex systems from a holistic perspective. The process should start with setting a holistic goal that ties together what people value most deeply in their lives with their life - supporting environment. Through a planning process within a holistic framework and by testing decisions against these values and the condition of the environment, people can consistently make better decisions for themselves and also the fauna, flora and environment on which all live depends.

Table 6: Controls used in Traditional Forest Management

	Basis of group rules	Example
1.	Harvest only selected components	✓ Tree: timber, fuelwood, fruit, nuts, seeds, honey, leaf fodder, fibre, leaf mulch, other minor products (gums, resins, dyes, liquor, plated leaves, etc. ✓ Grass: fodder, thatching, rope; ✓ Other wild plants: medicinal herbs, food, bamboos, etc. ✓ Other cultivated plants: maize, millet, wheat, potatoes, beans, vegetables, fruits, etc. ✓ Wildlife: animals, birds, bees, other insects, etc.
2.	Harvesting according to condition of product	✓ Stage of growth: maturity, alive or dead ✓ Size, shape ✓ Plant density, spacing ✓ Season (flowering, leaves fallen, etc) ✓ Part: branch, stem, shoot, flower
3.	Limiting amount of product	✓ By time: season, days, year, several years ✓ By quantity: of trees, head loads, baskets, animals ✓ By tools: sickles, saws, axes, hoes, cutlasses
4.	Using social means for protecting areas	✓ By watcher: paid in grains or cash ✓ By rotational guard duty ✓ By voluntary group action ✓ By making mandatory the use of herders to watch animals.

Source: Modified after Arnold and Campbell (1985)

This will require building human and social capital through combining training in holistic management with training related to village bank groups, forestry groups, apiculture, gardens, permaculture, game guiding, ecotourism skills, alterative income generating activities and water resource management and by demonstrating that the land can be profitably restored without compromising their values and livelihoods.

Acknowledgements

The paper acknowledges the contribution of Ewane Basil and Effange Emilia (Department of Geography, University of Buea) for the administration of questionnaires.

References

Amiet, J-L and Dowsett-Lemaire, F. (2000) Un nouveau Leptodactylodon de la Dorsale Camerounaise. (Amphibia, *Anura*). *Alytes* 18 : 1-14

Arnold, J. and Campbell, J. (1985) collective management of hill forest in Nepal; community forestry project. Washington DC. p. 7-15

Asong, A. (2001) Forest reserves and forest reserve strategies in South West Province. In: J. Dunlop and R. Williams (eds) *Culture and Environment*, University of Buea/University of Stratchdyde in Glasgow. p. 116-130.

Balgah, S. (2001) Exploitation and conservation of biological resources in Mount Cameroon region. In; C.M. Lambi and E.B. Eze (eds). *Readings in Geography*, Unique Printers, Bamena. p. 310-324.

Bawden, C, and Andrews, S. (1994) Mount Kupe and its Birds. *Bulletin of African Bird Club* 1:13-16.

Collar, J. and Stuart, S. (1988) Key forests for threatened birds in Africa. ICBP, Cambridge. p. 10-21

Denniston, D. (1995) Sustaining mountain people and environments. Worldwatch Institute. W.W. Norton and Company, London. p. 38-57.

Effange, E. (2006) The involvement of local population in protected area management in Cameroon; Case of Bambuko Forest Reserve and Bayang-Mbo Wildlife Sanctuary. Unpublished MSc. Thesis, Dept. of Geography, University of Buea. p. 60-78.

Ewane, B. (2006) Optimising the management of dynamic ecosystems. Case of the Southern Bakundu Forest Reserve. Unpublished MSc. Thesis, Dept. of Geography, University of Buea. p. 63-75

Dowsett, R. (1989) Preliminary natural history survey of Mambilla Plateau and some lowland forests of eastern Nigeria. Tauraco Research Report. No. 1. 56p.

Dowsett-Lemaire, F. and Dowsett, R. (1988) Zoological survey of small mammals, birds and frogs in Bakossi and Mount Kupe, Cameroon. Unpublished Report for WWF – Cameroon, 46p.

Dowsett-Lemaire, F. and Dowsett, R. (2000) Further biological surveys of Manenguba and Central Bakossi in March 2000, and an evaluation of the conservation importance of Manenguba, Kupe, Bakossi and Nlonako, with special reference to birds. Unpublished report for WWF – Cameroon, 45 p.

Fonyam, J. (2001) The legal protection of forests, a case study of Cameroon. In: J. Dunlop and R. William (eds.) *Culture and Environment*, University of Buea / University of Stratchdyde in Glasgow. p. 39-63.

Fotso, R.; Dowsett-Lemaire, F.; Dowsett, R. (2001) Cameroon Ornithological club, Birdlife Conservation Series No. 11, ICBP, Cambridge. p. 133-159

Gartland, S. (1989) La Conservation des écosystèmes forestiers du Cameroun. IUCN, Gland and Cambridge. p. 5-15

Jaff, B. (1994) Management of protected areas with particular attention to poaching and cross-border cooperation in the South West Province . *Unpublished paper presented in the Rgional Concertation on the Environment in Buea*, MINEF, Yaounde; March 1994. 10 p

Jongman, H. (1995) Nature conservation planning in Europ; developing ecological networks. *Landscape and Urban planning* 27: 253-258

Kamanda, B. (1994) The Southern Bakundu Forest Reserve Project: a rural economy survey. ITTO/ONADEF, Yaounde, p. 7-8.

Kelkit, A., Ozxel, A. and Demirel, O. (2005) A study of the Kazdagi (Mt. Ida) National Park: an ecological approach to the management of tourism. *Int. Journal of sustainable Development and World Ecology* 12: 1-8.

McNeely, J.; Miller, K.; Reid, W., Russel, A. and Werner, T. (1990) conservation the World's biological diversity. IUCN/World Bank, Washington D.C. p. 11-12.

Mesumbe, I. (2001) The ecology of Mount Kupe Forest. Unpublished Long Essay, Dept. of Geography, University of Buea. 46p.

Ndenecho, E. (2005) Conserving biodiversity in Africa. Wirldlife management in Cameroon. *Loyola Journal of Social Sciences* 19 (2): 211 – 228.

Scatterfield, A.; Crosby, M.; Long, A.; Wege, D. (1988) Endemic bird areas of the world; priorities for biodiversity conservation. Birdlife Conservation Series No. 7, Cambridge. p. 11-18.

Stuart, S.N. (1986) Conservation of Cameroon Mountain forests ICBP. Cambridge p. 12-19.

Tucker, C. (2000) Striving for sustainable forest management in Mexico and Honduras: the experience of two communities. *Mountain Research and Development* 20: 116-117

White, F. (1983) The vegetation of Africa, a descriptive memoir to accompany the UNESCO/AETFAT/UNSO vegetation Map of Africa (3 plates), Allard Blom, UNESCO, Paris. Plate 1, 2 and 3. At 1:5,000,000.

Chapter Six

Climographic Analysis and Mapping of the Mount Cameroon Region

Summary

Mountain environments possess fragile ecosystems which are particularly very sensitive to climate change. Mount Cameroon is the highest mountain peak (4095m) in West Africa and the foot slopes are a microcosm of tropical plantation agriculture. It also possesses a diverse and rich biodiversity of scientific and conservation interest. The area is therefore a region with much scientific, industrial and agricultural activity. The paper analyses, collates and maps rainfall, temperature and sunshine data using a combination of observation and secondary data. It maps and presents these weather elements in climographic form for easy access to the local scientific, industrial and farming communities. Such data is indispensable for land resource evaluations and the planning of development projects. The paper concludes that the existing network of weather stations is located around the base of the mountain in the zones under Cameroon Development Corporation (C.D.C.) agro-industrial enterprises. Available data are mainly of agricultural interest with no indication on altitudinal climatic variation. Reliable long-term monitoring is needed. The range of climate over the whole mountain undoubtedly has an influence over its biodiversity, but the relationships are complex and cannot be unraveled without more climate data. These are necessary for the effective management of the natural resources of the region.

Introduction

Mount Cameroon region is a microcosm of tropical tree crop agricultural development and the need for agro-meteorological data has since been at the centre of the activities of the Cameroon Development Corporation (C.D.C.). Since 1947, the C.D.C. has focused on research and agro-industrial development in the area. It has always made their weather data freely and generally available to the public. However, in recent years, particularly with the development of the University of Buea and its outreach research projects in the area, and the development of Limbe Botanic Garden, an increasing amount of scientific activity is being carried out in the region. The University of Buea and the Limbe Botanic Garden run weather stations in their premises that do not represent a region with many scientific and agro-industrial activities.

The aim of this study is to analyze and present the long and medium term rainfall, temperature and sunshine data in a cartographic form. More specifically, it seeks to present the available data in a form which is easily accessible and of value to the local scientific and industrial community.

The Study Area

The study area is located on the coast of the Gulf of Guinea in the South West Province of Cameroon. The mountain rises steeply, in as little as 18 km, from sea level to 4095m at the summit. The summit of the mountain is located at $9^{0}10$'E and $4^{0}13$'N. It is the highest mountain in West Africa. It has several unique characteristics in that it appears on a map as a single peak, almost completely ringed by the 100m contour line, when in fact it is very much the chain of the volcanoes which extend from Annabon and Principe islands in the Atlantic Ocean almost all the way to Lake Chad. As the base of the mountain lies at the sea level and the summit at 4095m above sea level, there is a climatic variation with aspect and altitude. The climate presents some unique characteristics. Local variations of rainfall and temperature are striking over short distances. The area has a rich and important biodiversity.

The variation in rainfall and solar radiation together with rich volcanic and alluvial soils has made the region a microcosm of tropical plantation agriculture. The largest agricultural undertaking is the agro-industrial plantations of the Cameroon Development Corporation (C.D.C.) In addition to this para-government corporation, there also exists the PALMOL Company (Unilever Limited, London) which operates estates in Meme – Ndian River Basin. The C.D.C. operates mainly in the mountain region. Both companies are expanding rapidly and sell their products directly on the world market. The main crops include oil palm, bananas, rubber, tea, pepper and coconuts. Both companies associate small-holder production with estate production. Other commercial crops include cocoa, robusta coffee, and a variety of tuber crops. In agro-industrial plantations, supplementary irrigation is used in nurseries and banana plantations. There is therefore a need for agro-meteorological data.

Table 1: Key geographical variables for centralize each site

Site	Altitude (m)	Distance from coast (km)	Direction from coast
Mokoko	200	21	East
Idenau	40	1	On the coast
Debundscha	20	0.5	On the coast
Mokundange	40	0.5	On the coast
Mabeta	20	20	In the mangrove creeks
Tole	700	25	North-east
Molyko	400	36	North-east
Mpundu	40	44	North-east
Malende	40	24	North-east
Mbonge	50	25	East-north-east

The C.D.C. maintains an extensive network of meteorological stations throughout their plantations in the region. Variables which have the greatest bearing on agriculture are monitored, though not all are available at each site. This study used a set of weather stations on the basis of their geographical distribution, variables recorded and duration and reliability of the recording (Table 1)

Methods And Data Sources

The climatic variables obtained from the C.D.C. meteorological service were rainfall, amount, rainfall days, sunshine and temperature. The data was obtained for a period of 30 years for most sites. These were complemented by the work of Fraser *et al.* (1998). An 80% + probability for maximum and minimum rainfall estimates (that is, that likely to occur in 8 out of 10 years) is considered to be useful in hydrology, agriculture and land use planning (Fraser *et al.*, 1998; Woodhead, 1982). Fraser *et al. (*1998) used the frequency distribution to predict maximum and minimum total rainfall likely in 8 out of 10 years (Table 2). The observed frequencies conformed to the expected normal distribution in only three of the eight sites (Chi-square-test).

Data for the remaining five sites were transformed using square root but still did not approximate to a normal curve. According to Fraser *et al.* (1998) at these five sites the magnitude of variation was such that longer runs of data were needed to make statistically significant predictions. The climographs presented in this study are derived from data collected from the reliable weather stations (Ndenecho, 1984; Fraser *et al.*, 1998; Stanford, 1968; F.A.O, 1977; Zogning, 1983).

Table 2: Total annual rainfall for each site

Site	Mean Rf (mm)	Max Rf (mm)	Min Rf (mm)	s.d.	Sample size (years)	Max expected in 8/10 years	Min expected in 8/10 years
Mokoko	2844	9709	1899	509	12	-	-
Idenau	8392	12449	3303	1866	30	9750	6710
Debundscha	9086	16965	4153	3792	39	-	-
Mokundange	4935	8327	1816	1438	28	6150	3720
Mabeta	4384	6791	1928	1040	30	-	-
Tole	2743	4978	1503	771	30	3291	2112
Molyko	2141	2867	1356	372	29	-	-
Mpundu	2085	5246	438	1033	27	-	-
Mbonge	2192	3102	1475	444	29	-	-

Results and Discussions

In the global context, the Mount Cameroon region has a tropical seasonal climate with local variations: areas located at the western edge of the mountain, areas exposed to the ocean and areas at the leeward side of the mountain. All the weather station however fit into the "tropical rainy" (A) category of the Köppen climate classification. Six of these major categories are used for the world. Category A is characterized by a coolest month with temperature 18^{0}C and above. As the passage of the sun varies throughout the year, the Inter-tropical Convergence Zone (ITCZ), where winds from the southern and northern hemispheres meet, shifts across the equator (Bradshaw and Weaver, 1992). As a result the region at different times of the year is under the influence of either continental winds from the north-east or maritime winds from the south-east.

The coastal location and bulk of the mountain interfere with the wider climatic pattern. Within the sample of climatic stations investigated round the mountain, there are elements of three different tropical climatic regimes: 1) equatorial, with rain all year round; 2) seasonal, with wet and dry periods and < 60 mm precipitation in the driest month; and 3) monsoon, with great contrast between seasons.

Climatic Pattern

The climate is monsoonal, that is to say, its feature include high yearly totals of rainfall with hot temperature, yet a marked dry season. The climate is governed by the interaction of the S.W. monsoon winds which are warm, moist, unstable winds from the Saint Helena Anticyclone and the dry, thirsty, dust laden; cool N.E. Trade Winds from the Sahara. The N.E. Trade Winds are stable and bring dry conditions while the S.W. Monsoon Winds are unstable and bring wet conditions. They meet along a front which moves north and south following the apparent movement of the sun. Rainfall is greatest along the line of discontinuity where they meet (Intertropical front). These broadly govern the climatic pattern which presents a rainy season and a dry season. The circulation near the surface is represented by the 850 millibar level, corresponding to about 1500 m above sea level. Since the mountain is 4095m in height, the influence of orographic factors, therefore, becomes very important in modifying the climate of the region.

Orographic influences on temperature

The general uniformity of temperature with place which prevails in the tropics is interrupted by one major factor – elevation. It causes large variations over short distances. Broadly, temperatures decrease with increasing altitude but the lapse rate is far from uniform. The main controlling factor is orographic clouds. Under cloudiness, when the effects of radiation are greatly reduced, the lapse rate tends to be similar to that of the free atmosphere which is usually around 5^0 C/100m. Still, the effects of cloudiness on lapse rates can not easily be generalized over a large area, because of its variability. Clouds develop over the highland during the day, mainly as a result of local convection and orographic lifting, while the lowlands remain clear.

Insolation characteristics

The daily duration of insolation is greater in the afternoon than in the morning; 55.5% and 44.5% respectively (Zogning, 1983). The actual variation or dispersion of annual insolation is determined by estimating the coefficient of variation in monthly sunshine hours at various locations (Table 3)

Table 3: Coefficient of variation in monthly sunshine hours

Station	Mean of 12 months (hours)	Standard deviation (hours)	Coefficient of variability (%)
Idenau	109	53	49
Bota	123	59	48
Ekona	103	44	43
Meanja	115	44	39
Kukonje	139	34	25
Tombel	100	41	41
Mbonge	118	43	36
Tole	79	43	55

Windward slope locations (Tole, Idenau, Bota) show a monthly high variability in sunshine hours. Lee slope inland locations (Meanja, Mukonje and Mbonge) have a lower variability. The higher variability of sunshine hours in Tombel may be attributed to the orographic effects of Mount Kupe. On the whole the region experiences very low diurnal sunshine hours (Table 4). They are particularly low during the rainy season. This is attributed to the dominance of orographic clouds following the orographic uplift of monsoon winds. Orographic fog and mist is also persistent during the wet season. Conversely, during the dry season the dust laden harmattan or N.E Trade Winds only permit filtered sunshine. Tole at an elevation of 731m on the windward side has extremely low diurnal sunshine hours. Lee slope locations such as Meanja and Mukonje have higher values. It is possible to observe 2 to 3 months of consecutive absence of sunshine in Tole. In 1981 only 15.4 sunshine hours were recorded in 90 days (1st of July to 28th of September). In 1982 this dropped to 4.5 hours during the same period.

Table 4: Average daily sunshine hours (hours/day)

Stations	Months/ years	J	F	M	A	M	J	J	A	S	O	N	D
Idenau	9	5.2	6.0	4.7	5.0	4.8	2.6	1.6	1.0	0.8	2.4	3.5	4.5
Bota	7	6.4	5.8	4.9	5.5	4.2	2.2	1.2	1.1	2.1	3.4	5.3	5.8
Ekona	9	4.5	4.8	4.0	7.8	4.3	2.7	1.1	1.0	1.6	3.2	4.1	4.3
Meanja	7	5.3	5.2	4.3	5.1	5.0	3.2	1.7	1.2	1.8	3.4	4.3	4.5
Kukonje	7	5.8	5.4	5.4	5.6	5.8	4.2	2.8	2.9	3.2	4.2	4.7	4.4
Tombel	9	3.9	5.5	3.7	4.1	4.1	2.2	1.5	1.2	1.8	2.7	3.8	4.5
Mbonge	11	5.1	5.7	4.7	4.5	4.8	3.7	1.8	1.8	1.1	3.0	4.4	5.2
Tole	16	4.5	4.1	3.2	4.1	3.0	1.7	0.7	0.6	0.8	1.9	2.8	3.5

Characteristics of cloudiness

Information about cloudiness has been obtained by the observation of actual hours of sunshine. The simple formula S+C = 100, in which S is the actual sunshine as a percentage of its maximum possible duration, and C is cloudiness, expressed as a percentage of the visible sky (Niewolt, 1977). This gives a reliable approximation but all data obtained are for day time only.

From Table 5 cloudiness varies strongly at short distances with respect both to location and season. There seems to be some correlation between the distribution of cloudiness and rainfall. This is probably due to the occurrence of intense orographic clouds during the wet season. The lee locations have lower cloudiness values. This is logical due to the subsidence of air movement at this lee zone at the dry adiabatic lapse rate (DALR)

Frequency of rainfall

The frequency of rainfall is indicated by the number of rain-days (Fig. 1, 2, 3 and 4). There is a strong correlation between the number of rain-days and the total amount of rainfall. The number of rain days decrease with elevation on the windward slopes. Conversely, at the lee slopes rain-days decrease with decreasing elevation and increasing continentality except the Tombel-Mukonje area where Mount Kupe orographic influences prevail. The windward slopes show a more regular frequency of rainfall. About 18-20 days experience rainfall during the dry season on the windward locations as opposed to less than 13 rain-days on the lee locations. Bota to the south east of the region has such low values as well.

Table 5: Cloudiness (c) expressed as a percentage of the visible sky

Stations	Months / years	J	F	M	A	M	J	J	A	S	O	N	D
Bota	7	46.7	48.5	59.4	64.7	64.7	81.3	90.0	90.8	83.6	69.3	56.3	51.6
Ekona	9	62.9	57.1	66.9	63.9	63.9	77.2	90.5	91.9	86.9	73.6	65.5	64.5
Meanja	7	56.1	54.0	63..9	58.0	58.0	73.0	85.4	89.2	85.2	72.0	64.4	62.9
Kukonje	7	53.4	52.0	55.1	51.8	51.8	65.0	76.8	75.8	73.0	65.0	60.5	62.9
Tombel	9	67.2	52.0	69.3	66.1	66.1	81.6	87.3	89.2	85.0	77.1	68.3	63.4
Mbonge	11	58.0	49.1	61.0	62.5	60.2	69.4	84.9	85.2	86.1	75.2	63.3	56.9
Tole	16	62.9	63.7	72.8	65.8	75.0	85.5	94.3	94.3	93.6	83.8	76.3	70.4
Idenau	9	56.4	46.2	61.0	58.3	60.2	78.6	86.8	91.9	89.4	79.8	70.5	62.6

Rainfall intensity

Rainfall intensity is estimated by calculating the mean rainfall amount per rain-day. The results are presented in Table 6.

Stations	Rainy season	Dry season	Annual
	July-September (mm/day)	Decmember-February (mm/day)	
Mbonge	26	10	24.6
Idenau	54	20	48.5
Debundscha	52	46	51.2
Mokondange	43	12	36.3
Bota	32	18	30.3
Moliwe	30	11	27.7
Tole	25	12	23.0
Molyko	17	12	16.3
Tiko	19	11	18.0
Ekona	17	13	16.6
Meanji	17	10	16.2
Mondoni	20	10	19.2
Missellele	22	8	20.6
Mukonje	16	10	15.5
Tombel	25	17	23.8

Striking variability in mean rainfall per rain-day occurs in time and space. Generally, it increases with the total amount of rainfall as seen in the windward slope locations. (Idenau, Debundscha, Mokondange, Bota, Tole). The mean number of rain-days per year follows this pattern (Fig. 3) Orographic uplifting is probably the main factor responsible for these differences. Generally, intensity (volume of rainfall per unit time) increases with the total amount of rainfall, and the only exceptions to this rule are found in the highlands at levels over 1500m, where the number of rain-days increases with higher elevations while the rainfall volume reduces. This reduces the mean rainfall per rain-day. The second statement seems to hold more for the lee slopes than the windward slopes.

The variation between wet and dry season rainfall is greatest at the coastal sites (Fig. 5-11) which have a characteristically monsoon precipitation regime. These high rates of rainfall at the coastal sites are clearly the result of orographic precipitation as moisture laden oceanic air is forced to rise in a short distance up and over the slopes of the mountain. Debundscha has the highest mean annual rainfall (Fig. 7). It is on the Coast and is the closest station to the southwestern slopes of the mountain. Here air borne by the south-westerly onshore wind is forced upwards over a very short distance, consequently unloading moisture over a small area of land. Because of its southerly aspect, the prevailing wind at Mokundange passes along the coast rather than coming directly from the sea, thereby passing around and over the flanks of the mountain depositing rain over a wider area (Fig. 8). Maritime air passing over Mabeta, further down the coast, is not being forced upward, hence the less intense and lower rainfall there (Fig. 9). Molyko, Malende (Fig. 11) is very much in the rain shadow of the mountain and this is reflected in their maximum wet season rainfall being less than a quarter of that at Debundscha and Idenau (Fig. 6).

Fig. 1: Mean Annual Rainfall in the Mt. Cameroon Region

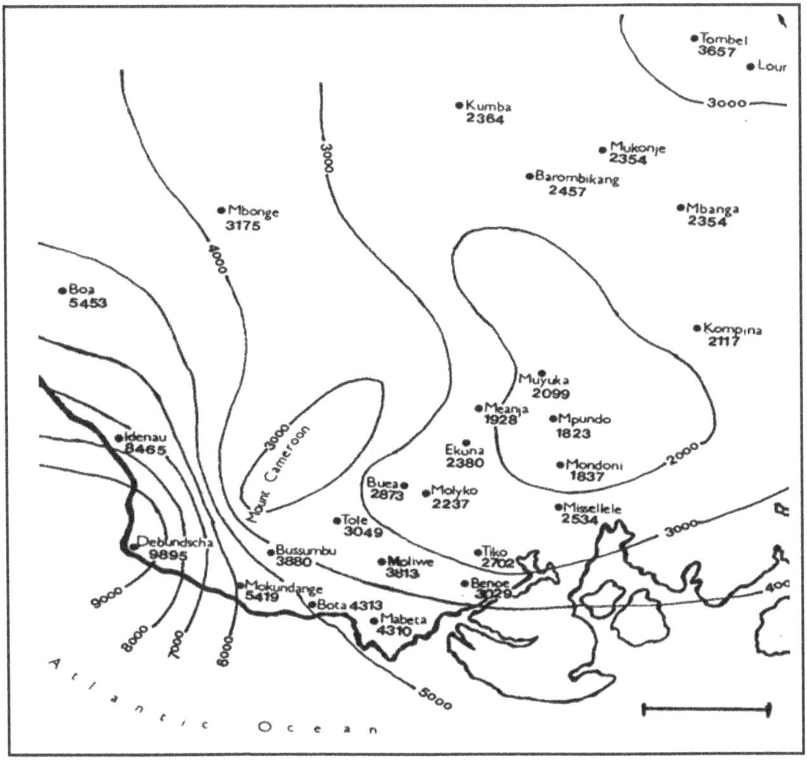

Fig. 2: Mean Annual Number of Rainy Days in the Mount Cameroon Region

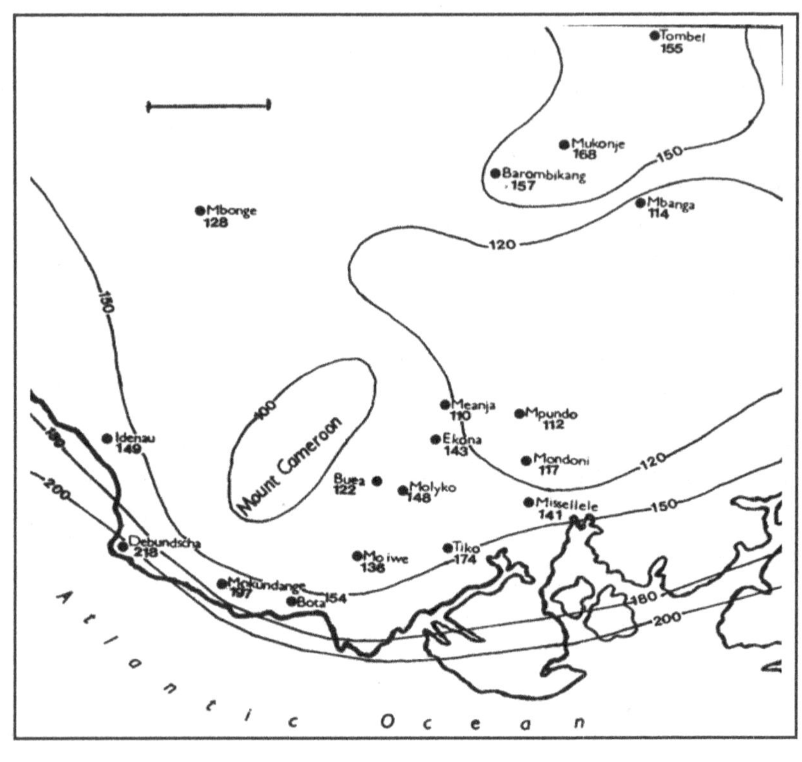

Fig. 3: Mean Rainfall and Number of Rainy Days During the Season in the Mount Cameroon Region (December-January-February)

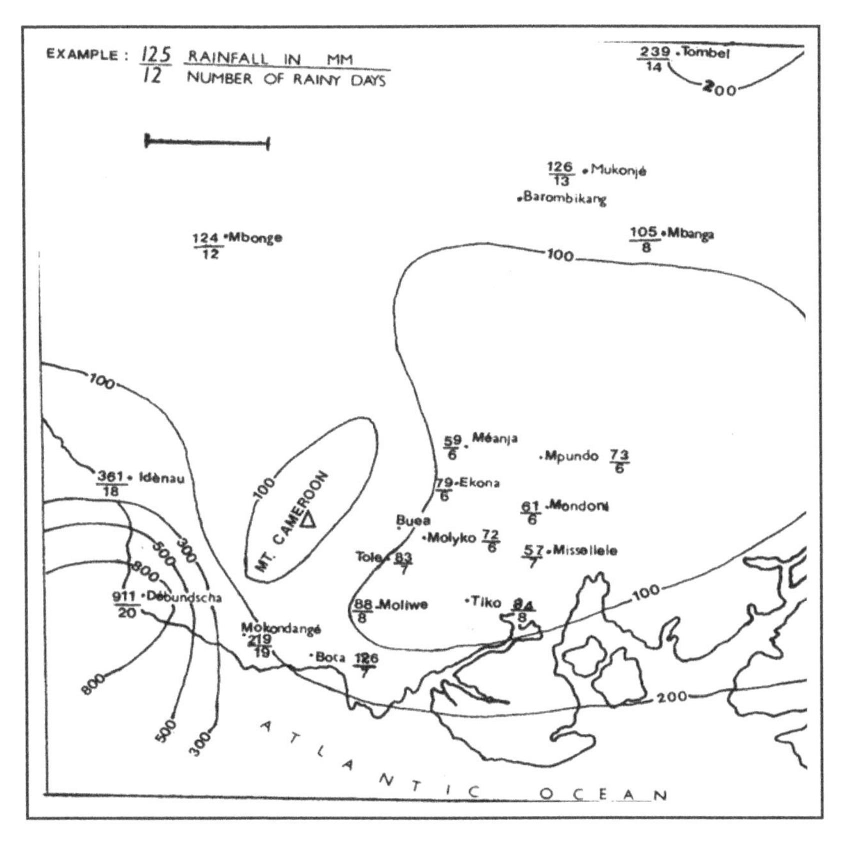

Fig. 4: Mean Rainfall and Number of Rainy Days During the Rainy Season (July-August-September) in the Mount Cameroon Region

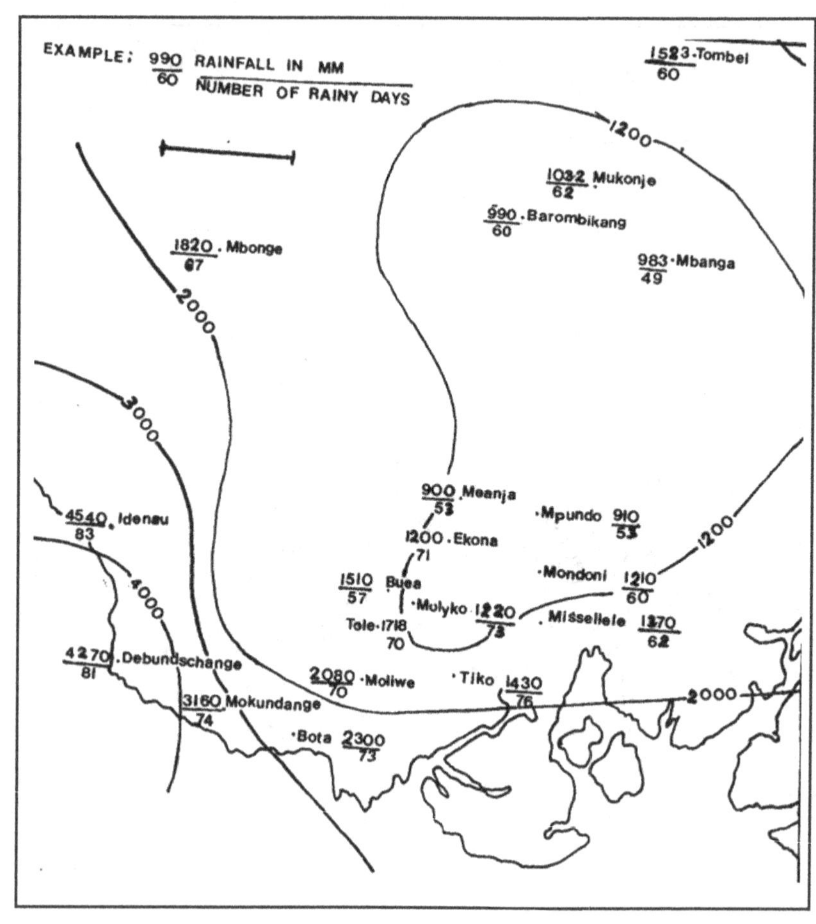

Fig. 5: Climographs, Mokoko and Idenau: Mean Maximum, and Mean Minimum Temperature; Mean, Absolute Maximum and Absolute Minimum Sunshine Hours, Rain Days, And Mean Total Rainfall

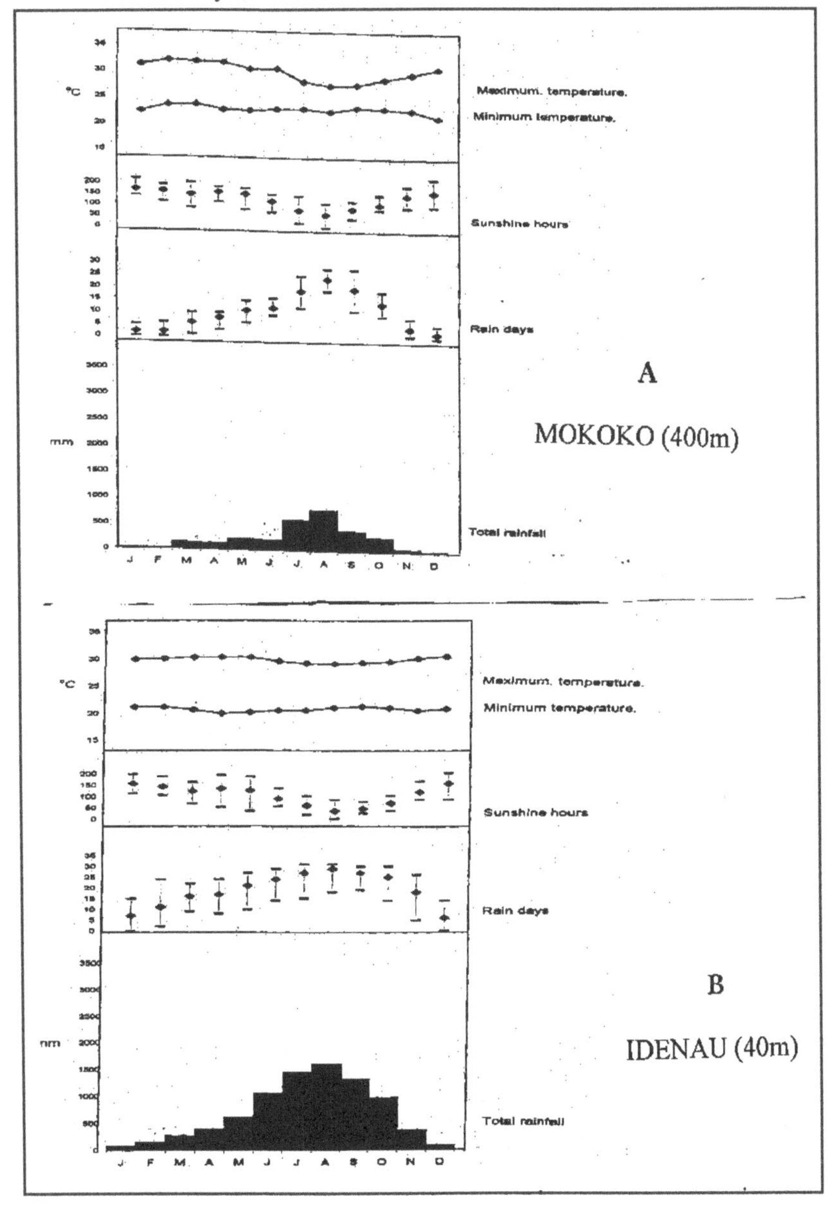

Fig. 6: Climograph for Debuncha and Mokundange: Mean maximum and mean minimum temperature; mean, absolute maximum and absolute minimum sunshine hours, rain days, and mean total rainfall

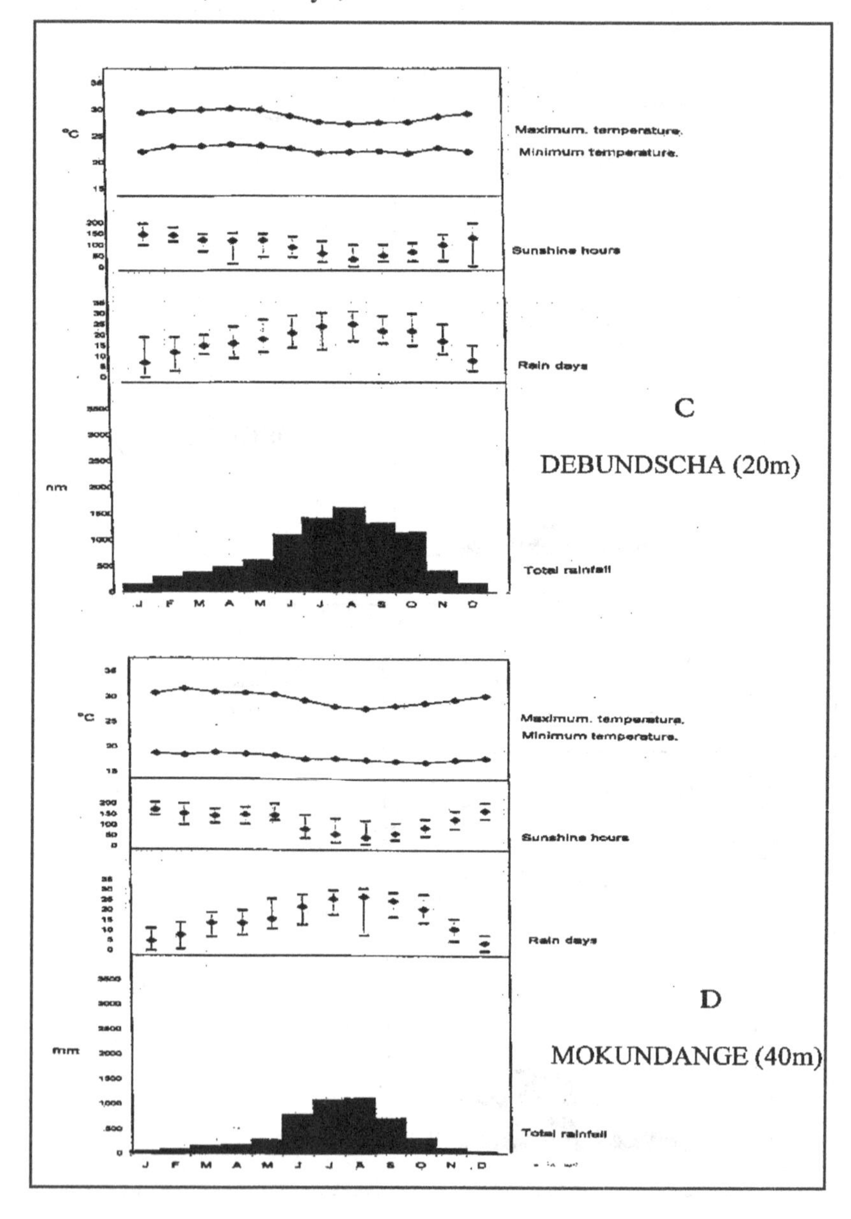

Fig. 7: Climograph, Mabeta and Tole: mean maximum, and mean minimum temperature; mean, absolute maximum and absolute minimum sunshine hours, rain days, and mean total rainfall

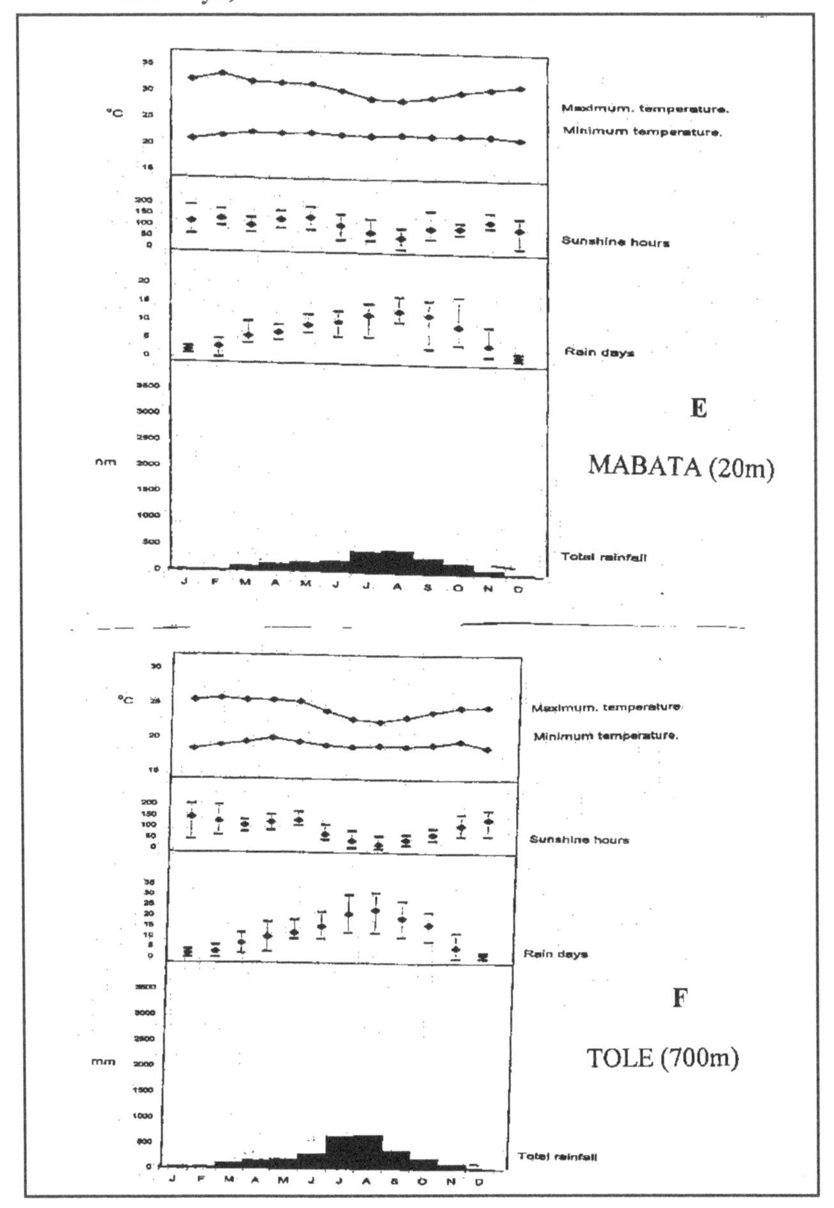

Fig. 8: Climograph for Malende and Mpundu: Mean maximum and mean minimum temperature; mean, absolute maximum and absolute minimum sunshine hours, rain days, and mean total rainfall

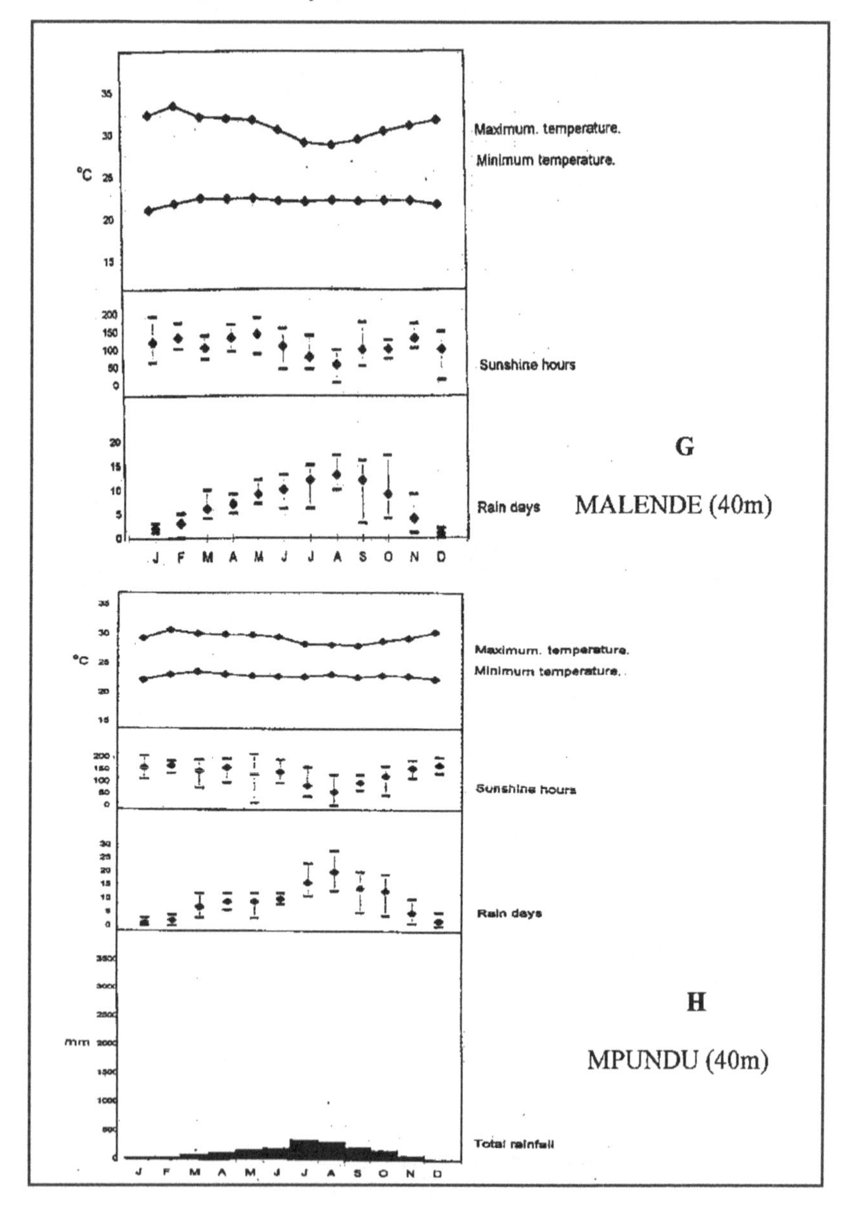

Conclusion

The existing network of weather stations is located around the base of the mountain in the zone presently used by CDC for plantation agriculture. The data from these stations provides no indication of altitudinal climatic variation. Some high altitude data exists from German colonial times; reliable, long-term monitoring stations are needed. The range of climate over the whole mountain undoubtedly has an influence on its great biodiversity, but the relationships are complex and cannot be unravelled without more climate data. This knowledge is important if we are to fully understand, and therefore effectively manage, the natural resources of Mount Cameroon.

The first indications of climate change are often most obvious in mountain environments. Species inhabiting mountain environments are particularly sensitive to environmental changes: the plant life is considered to be growing at its limit in terms of resource availability and environmental stress. A network of weather stations at a range of altitudes on Mount Cameroon would provide valuable climate data to feed into global climate change models.

References

Bradshaw, M and Weaver, R. (1993) Physical geography: an introduction to earth environments. Mosby, USA

F.A.O. (1977) Soil surveys and land evaluation for the second development programme of the CDC. *Soil Science Project, Technical Report No. 7*, Bota.

Fraser, P.; Hall, J. and Healey, J. (1998) Climate of Mount Cameroon region. Mount Cameroon Project, Limbe Botanic Garden, Limbe.

Ndenecho, E. (1984) Oil palm production in the South West Cameroon: a review of the industry's performance. Unpublished MSc. Thesis, Silsoe College, England.

Neuwolt, S. (1977) Tropical climatology. John Wiley and Sons, New York.

Standford, H. (1968) The Cameroon Development Corporation: partner in national growth. Bota, West Cameroon

Woodhead, T. (1982) Variety of seasonal and annual rainfall totals in East Africa. *East African Agricultural and Forestry Journal, vol, 45*, p. 74-82.

Zogning, A. (1983) Note sur l'inclinaison des héveas dans la region du Mont Cameroun. *Cameroon Geographical Review vol. 4, No. 2*, Yaounde University.

Chapter Seven

Ecological Planning and the Potential for the Development of Ecotourism in Kimbi Game Reserve, Cameroon

Summary

Game reserves and other protected areas are potential areas for the development of ecotourism because of their biodiversity, landscapes and cultural heritage of local or indigenous people. This study investigates the environmental sustainability of game reserves using a sample of the Kimbi Game Reserve. It assesses the potentials of the reserve for the development of ecotourism by employing a combination of field observation, examination, data collection and evaluation, using a SWOT analysis. The SWOT analysis determines opportunities and threats, and strategic suggestions for ecological planning. The study determines usage potential and the types of ecotourism feasible for development, and appraises the current management strategies. It concludes that ill-adapted strategies are bound to fail in promoting ecotourism, attaining sustainable landscapes and livelihoods. The Kimbi Game Reserve has major economic potential for ecotourism which can be realized by integrating the cultural values, livelihoods and environmental awareness of local people in tourism development. Finally, the paper recommends that in this process, government organizations, universities and research institutions must interact sufficiently in order to develop the potential of interest to ecotourism, ecocultural tourism and scientific tourism.

Introduction

The main focus of most interpretations of sustainable development is the reorientation of understanding society in relation to nature (Redclift, 2000). Even though this focus does not necessarily imply such an outcome, most implementation of sustainability has been satisfied with the integration of environmentally sound practices and policies into development programmes and projects (Chifos, 2006; Lele, 1991). Building on the foundation of increasing environmental awareness, interpretation of economic development as integral to environmental and social systems has gained momentum and has been expressed in a variety of ways, such as redesign of economic processes to work with nature instead of against it (McDonought and Braungart, 2002) or rethinking the linkages among livelihood strategies, poverty alleviation and environment (Chambers, 1992; Neefjes, 2000). Thus, the physical and biological, environmental and economic components of the world system are firmly ingrained in the interpretation and operationalisation of sustainable development.

Protected areas in several countries have been damaged when important ecological aspects of such areas have not been considered. In these areas, plans based on ecological data are needed for land use planning, improvement and development (Jurgen, 1993). In most developing countries the purpose of protected areas is to conserve biodiversity and so have failed to recognize the realities of their local socio-cultural and economic environments (Ndenecho, 2007). According to Ndenecho (2005) they must protect the cultural, natural and traditional activities of people against the consequences of rapid progress. Effective plans need to conduct all relevant biological, social, physical and economic factors and focus on important resources affecting the ecological integrity of the areas (Sanderson *et al.*, 2002; Gengiz, 2006).

Several studies conducted in protected areas have focused on ecotourism or nature tourism as a form of sustainable tourism (Poiani *et al.*, 1998; Daniel *et al.*, 2005). Recently, research has focused on how protection of local ethnicity can be achieved without impacting on the life of local people by linking social life and environmental protection (Barkin, 1996; Gregory, 2005). Ecotourism has been suggested as a key to sustainable development of protected areas (Barkin, 1996). It provides investment for tourism and enhances the living standards of local people by providing opportunities for employment. Cultural investments, such as historic preservation or dissemination of traditional skills, can also work to provide economic benefits while preserving connectivity with the past (Chifos, 2006), that is, it is nurtured and disseminated as poverty is alleviated.

The general trend in ecotourism is to increase experiences by encouraging activities such as long-distance walking, camping, boating, hunting, sight-seeing, swimming, cultural activities, bicycling, observing wildlife and nature, skiing, visiting historical places, and horse riding among others. Generally, instructive activities, for example, wildlife observation, participation in festivals, cultural activities and nature landscapes, attract most attention (Gengiz, 2007). In this study, the potential for the sustainable development of ecotourism is assessed in a sample of the Kimbi Game Reserve and the adjacent rural communities. This reserve is rich in biological, geographical and cultural values that are yet to be fully developed to attract local people and foreign tourists. The study was designed to determine the potential use of the area and to suggest ecotourism types likely to be beneficial for local people. It stresses the need for ecological planning and the linking of livelihoods with environmental protection projects.

The Study Area

The grid reference of the study area is Latitudes 6°5'N and 6°40N, and longitudes 10° 19'E and 10° 24'E. The total land surface area covered by the reserve is 6000 ha (Fig. 1). There are 13 villages around the reserve. The altitudinal range is between 950 and 1500m above sea level. Hawkins and Brunt (1995) have described the climate as a "sub-montane cool and misty climate" with an annual mean maximum temperature of 20°C to 22°C and mean minimum of 13°C to 14°C. Annual rainfall varies between 1780 mm and 2290 mm. Most of the rainfall occurs between July and September. A dry season occurs from mid – October to mid- March.

Geographically the area is part of the Cameroonian Highlands ecoregion which encompasses the mountains and highland areas of the border region between Nigeria and Cameroon (Stuart, 1986). The area falls within the Afromontane archipelago-like regional centre of endemism that spans the entire African continent. The forests in the area are refugia in montane and sub-montane environments. These are islands of biodiversity within anthropogenically degraded cultural landscapes. They support important local livelihoods and the socio-cultural activities of mountain dwellers. Conservation efforts have tended to emphasize the protection of biodiversity and so have ignored local livelihoods.

Fig. 1: Location Map of the Study Area Showing the Vegetation Types and the Population Size of the Main Chiefdoms

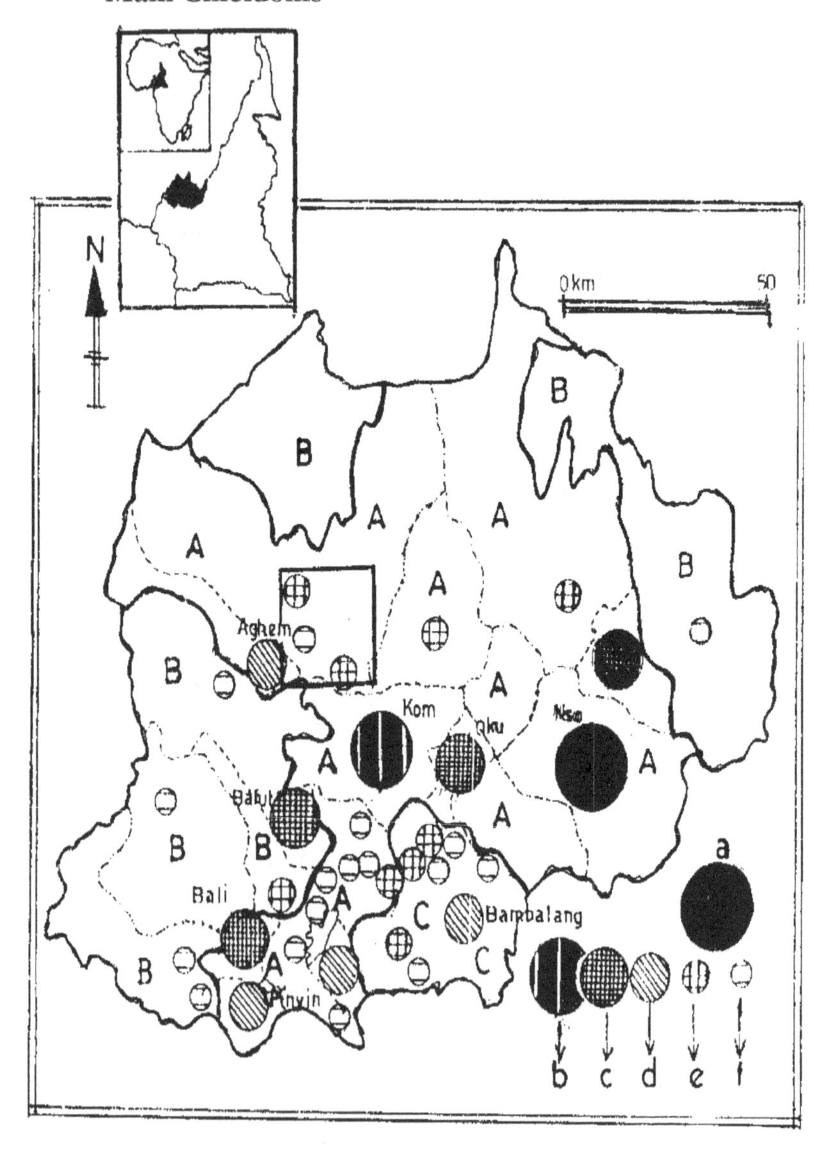

Legend: Population size: a) over 110.000; b) 72.000; c) 2000-35.000; d) 16.000-20.000; e) 1.000-15.000; f) 5.000-10.000 inhabitants

The montane forests are of great ecological significance. They contain several endangered species of plants and animals (Alpert, 1993; Ngwabuh, 2002). The forests provide local employment and livelihood. The area is spectacularly beautiful. Several volcanic episodes have created crater lakes, maars and strombolian cones. The reserve is in the Nyos volcanic District, with active gas eruptions and several thermo-mineral springs. The association of fissural and strombolian eruptions produced numerous spatter cones which give outstanding panoramas over the rift basins, forested valleys, and rugged grasslands. The magnificent views, unique wildlife and rich culture all have great tourist potential which could be realized with careful development. Tourists and scientists have shown interest in the wildlife and in the seismic processes of the area.

Climax vegetation and derivatives

Map code	Climax vegetation depicted in sacred forest groves	Anthropic derivatives within the chiefdom
A	Moist montane forest	Montane woodland, Tree Savannah, Shrub Savannah, Grass Savannah
B	Moist evergreen forest	Savannah woodland, Tree Savannah, Shrub Savannah and Grass Savannah
C	Moist evergreen forest and swamp forest	Tree Savannah

The game reserve was created in 1963 with the objective of promoting tourism and the improvement of the socio-economic development of the local communities. The reserve status has since remained on paper with little infrastructural investments and promotion of local livelihoods. In recent years it has come under serious threats from geological processes such as lake-basin gas eruptions and anthropic activities such as bush fires, agricultural encroachment, poaching and invasion by cattle during the dry season.

Research Methods

The study focused on the Nyos volcanic district where the Kimbi Game Reserve is located. Three potential resources for ecotourism were investigated. These were the natural landscapes, wildlife and cultural treasures. Base maps and aerial photographs produced by the National Geographic Institute were interpreted and updated by field verifications, and the use of archival material of the Kimbi Game Reserve. The cultural values of the people were assessed using informal interviews. Five randomly selected villagers were interviewed in seven villages within the vicinity of the reserve and the field assistance of the conservator. For the three aspects investigated, the study methods included observation, examination, data collection and evaluation, using a SWOT analysis, which is, determining the strengths (S), weaknesses (W), Opportunities (O) and threats (T) with regards to ecotourism development. This enabled the design of strategic suggestions for ecological planning and management.

Results And Discussions

As the world shrinks, tourists look for new destinations and new experiences. The Cameroonian Highland ecoregion offers both, with fascinating flora and fauna and new climates and terrains. The destinations and experiences in the area include nature landscapes, fascinating wildlife and a rich cultural heritage.

New Landscapes

The area is broken into plateau, fault blocks and steep escarpments. Several recent volcanic episodes have created strombolian cones, crater lakes and maars. Three types of eruptions were identified:

- Fissural eruptions are widespread with pockets of basalts at Aqulli, Machelli and Kuk villages. Some remnants are found on hill summits in the Nyos Volcanic district. They are extensive in the Nyos-Bwabwa area. In the Kam-Nyos valley they accumulated to form a volcanic-dammed lake, that is, Lake Njupi.
- Strombolian eruptions are common. Thirteen strombolian cones are found in the Nyos volcanic district at the vicinity of the game reserve. The cones have heights ranging from 100m to 150m above the local lowlands in Wum, Machelli

and Nyos village. Small cones or spatter cones are abundant and probably result from the association of fissural and strombolian eruptions.

- Phreatomagmatic gas eruptions are experienced in the Nyos volcanic district. On the 21st August 1986 a cloud of gas erupted from crater lake Nyos. The explosion killed 1700 people and 4000 heads of cattle. The Nyos volcanic district has since attracted international scientists and scholars interested in seismic processes. The crater lake basins liable to these processes include Njupi, Elum, Enep, Wum, and Nyi (Table 1). These are nested in depressions with sub-vertical slopes with the surroundings covered by pyroclastic deposits. These are probably air-fall deposits and basal surge deposits (streaming out of magma through the local rocks).

Fig. 2: The Kimbi Game Reserve and Feasible Zones for the Development of Ecotourism

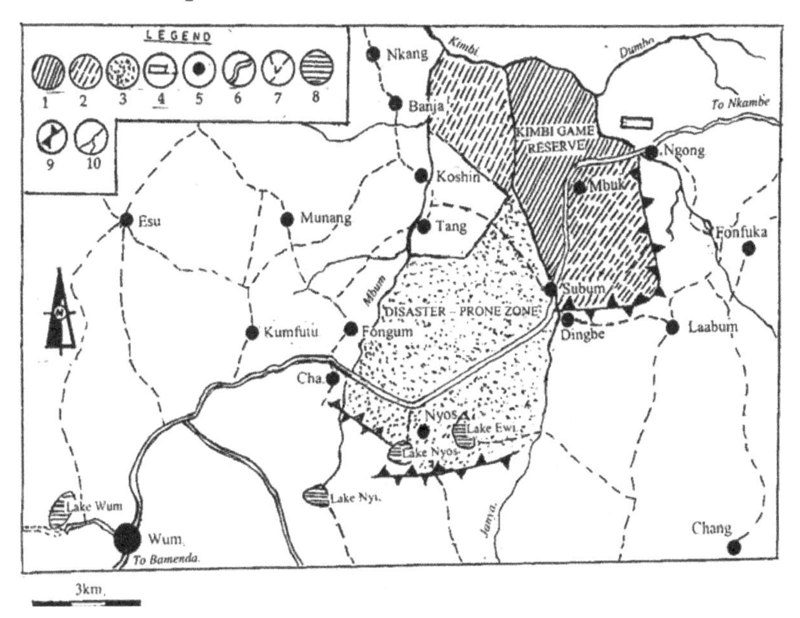

Legend: 1: Kimbi Game Reserve, 2: Picturesque Zone Feasible For the Extension of the Reserve, 3: Picturesque Toxic Gas Eruption-Prone Zone Feasible For the Extension of the Reserve, 4: Office and Residence of the Conservator, 5: Villages, 6: Earth Road, 7: Major Footpaths, 8: Lakes, 9: Mountain Divide, 10: Rivers.

Table 1: Lakes South of the Kimbi Game Reserve in the Nyos Volcanic District

	Lakes	Latitude	Longitude	Altitude (m)	Area (hectares)	Depth (metres)
1.	Enepe	6°18` N	10°02` E	697	50	78
2	Wum	6°24` N	10°03` E	1177	45	124
3	Njupi	6°25` N	10°18` E	1020	30	?
4	Nyos	6°26` N	10°18` E	1091	158	208
5	Elum	6°20` N	10°02` E	950	50	35
6	Bénakuma	6°26` N	9°02` E	576	154	132

The slopes are littered with landslide scars and some thermo-mineral springs. The physical and chemical properties of these sources may be beneficial for medical purposes, but there are no commercial facilities for using their water. The lakes have not also been developed for commercial purposes. This volcanic district which in 1986 was mapped out as a disaster zone and the villagers re-located lies to the south of the game reserve. Its fascinating landscapes and grass savannahs could be carefully developed as part of the reserve (Fig. 2).

Wildlife Resources

Wildlife is protected in the Kimbi Game Reserve. It has important potential in terms of flora which has attracted the interest of the International Union for the Conservation of Nature and Natural Resources (ICUN), The World Wildlife Fund (WWF) and the International Council for Bird Preservation (ICBP). It has been identified as one of the most at risk terrestrial ecological regions (Macleod, 1986; Stuart, 1986; Alpert, 1993). A total of 98 plants are recorded in the reserve (Table 2). These belong to 43 plant families (Kwanga, 2006). The ecoregion as a whole has one of the highest levels of endemism in the whole of Africa, particularly among birds and vascular plants. For example, 20 bird species are found only in this ecoregion (Stuart, 1986).

Table 2: Checklist of Plant Families and the Number of Plant Species in Kimbi Game Reserve

S/N	Family	Number of species
1	*Acanthaceae*	3
2	*Anarcadiacea*	2
3	*Annonaceae*	1
4	*Apocynoceae*	3
5	*Araliaceae*	1
6	*Bignoniaceae*	1
7	*Burseraceae*	1
8	*Combretaceae*	1
9	*Commelinaceae*	1
10	*Compsitae*	2
11	*Costaceae*	1
12	*Cyperaceae*	1
13	*Dracaenaceae*	1
14	*Euphorbiaceae*	6
15	*Graminae*	2
16	*Guttiferae*	1
17	*Lauraceae*	3
18	*Leganiaceae*	14
19	*Malastomocaceae*	1
20	*Meliaceae*	1
21	*Monispermaceae*	2
22	*Moraceae*	2
23	*Musaceae*	1
24	*Myristicaceae*	1
25	*Moraceae*	5
26	*Musaceae*	1
27	*Myristicaceae*	1
28	*Myrtaceae*	1
29	*Ochnaceae*	3
30	*Olaceae*	2
31	*Orchidaceae*	9
32	*Palmae*	2
33	*Piperaceae*	2
34	*Pihosporaceae*	1
35	*Rhamnaceae*	1
36	*Rosaceae*	1
37	*Rubiaceae*	9
38	*Sapindaceae*	1
39	*Ulmaceae*	1
40	*Urticaceae*	1
41	*Verbenaceae*	1
42	*Zingiberaceae*	2
43	*Pendenceae*	1
Total Number of Species		**98**

Table 3: The mammal population recorded in Kimbi Game Reserve

No.	Family	English name	Scientific name
1	*Colobidae*	Black and white colobus	*Colobus guereza*
2	*Cercopthccidae*	Olive baboons	*Papio anubis*
3	*Cercipithecinae*	patas monkey	*Cercopithecus (Erythrocebus) patas.*
4	*Cercipithecinae*	Green monkey	*Cercopithecus (caethiop) tantalus*
5	*Cercipithecinae*	Mona monkey	*Cercopithecus (mona) mona*
6	*Lagosmopha/ baridae*	Scrub hare	*Lepus scxtilis*
7	*Rodential/ sciuridae*	Striped ground squirrel	*Euxerus enythropus*
8	*Rodential/ sciuridae*	Red legged sun squirrel	*Heliosciurus rufobrackium*
9	*Rodential/ sciuridae*	African gaint squirrel	*Protoxerus stergeri*
10	*Hytrilidae*	crested porcupine	*Hystrix cristata*
11	*Thyronomyidae*	marsh canerat	*Thryonomys gregorianus*
12	*Cricetomyinae*	Emin's Giant Rat	*Cricetomys Emin's*
13	*Herspestidae*	Slender mongoose	*Herpestes sanguinea*
14	*Viverridae*	African civet	*Civettictis civetta*
15	*Procavidae*	Rock hyrax	*Procavia johustoni*
16	*Bivini*	African Buffalo	*Syncerus caffer*
17	*Tragelaphini*	Bushbuck	*Tragelaphus scriptus*
18	*Cephalophini*	Bay duiker	*Cephalophus dorsalis*
19	*Cephelophini*	Blue duiker	*Cephalophus monticola*
20	*Reduncini*	Rob	*Kobus kob*
21	*Reduncini*	Water Buck	*Kobus ellipsiprymus defassa*

Source: 2007 field work

Table 3 presents the 21 mammals recorded in the reserve. The population drastically reduced since 1973. It was reported that the once abundant species are rarely sited. This is attributed to grazing and farming encroachment, poaching, fire encroachment and poor implementation and management of the reserve status despite the scientific importance of the flora and fauna. A total of 203 bird species are recorded in the reserve (Table 4). These include 45 of the 215 Guino-Congo forest biome bird species and 8 of the 45 bird species restricted to the Sudan-Guinea Savannah. Species of interest include the brown-chested plover *(Venellus supercilious)*. This is an uncommon and local intra-African migrant found to breed in Cameroon and Nigeria (Alpert, 1993).

Povel's illadopsis *(Illadopsis puveli)* is an uncommon resident in the northern part of the reserve and outlier in savannah flock of 6 birds mostly observed in gallery forests in the Jonja River Valley. The bird life is threatened with habitat loss and fragmentation.

The main attractions are the lakes, the dark green forest of the reserve, gallery forests in valleys, natural pastures on the surrounding plains and hills, and the seasonal colours of plants. These create attractive views. The valleys formed by streams and hills enhance the visual value of the area, and topographic structure is an important feature in creating viewpoints. Present day geomorphic processes and the rich and unique biodiversity are of scientific importance and constitute important destinations for scientific tourism.

Table 4: Checklist of birds in Kimbi Game Reserve

Non-Passerines		Passerines	
Family	**No. of species**	**Family**	**No. of species**
Ardeidae	1	*Campephagidae*	2
Anatidae	1	*Crvidae*	1
Accipttridae	14	*Dicruridae*	1
Apodidae	3	*Eurylaidae*	1
Alcedinidae	8	*Emberizidae*	1
Charadrudae	1	*Estrildae*	10
Columbidae	4	*Fringillidae*	1
Cuculidae	8	*Hirundinidae*	7
Coludae	1	*Landidae*	2
Capitonidae	8	*Motacillidae*	4
Caprimulgidae	1	*Monarchidae*	2
Bucerotidae	1	*Muscicapioae*	3
Falconidae	1	*Malaconotidae*	10
Indicatoridae	1	*Nectarinidae*	11
Jacanidae	1	*Oriolidae*	2
Musophagidae	2	*Pycronotidae*	10

Table 4: Checklist of birds in Kimbi Game Reserve

Non-Passerines		Passerines	
Family	**No. of species**	**Family**	**No. of species**
Meropidae	4	*Platysteiridae*	3
Osittacidae	2	*Paridae*	1
Phasianidae	2	*Passeridae*	1
Picidae	3	*Ploceidae*	6
Rallidae	3	*Syviidae*	24
Scopidae	1	*Sturnidae*	7
Strigdae	1	*Thurdidae*	6
Threskiornithidae	1	*Timalinidae*	2
Upupidae	1	*Zosteropidae*	1
Total	**74**	**Total**	**129**
GRAND TOTAL = 203 species of birds			

Source: Summarized from the archival materials of the Kimbi Game Reserve

Cultural Landscape

The villages at the periphery of the reserve and the disaster – prone zone give the area an added attraction of traditional village life, in terms of settlement and variations in land use. The mountain pastures are colonized by semi-normadic tribes who keep cattle, sheep, and horses. Farmers settle in village chiefdoms on plains and valleys (Fig. 2). Traditional lifestyle and culture are important and attractive elements for tourists. Traditional architecture is an important element of the cultural landscape, with buildings of wood, bamboo and mud walls and grass-thatched, high, pyramid-like roofs. Numerous carved houseposts support the heavy-thatched roof with its comparatively wide overhang. Door-frames and door-=surrounds are carved with several symbolic motifs. Geological processes, configuration of the land, climate and biogeography have shaped the selection of buildings and construction materials. Festivals providing recreation are important in maintaining social cohesion in the chiefdoms. Annual festivals and dances in palace chiefdoms are attractions during the dry season. Tourists experience a variety of traditional meals and rites.

The reserve has no accommodation for visitors, a poor and seasonal road network and poorly developed sites and tourist infrastructure in the nearest town (Wum). These facilities are clearly inadequate for the accommodation needs of tourists. Handicraft production which is already on the decline (Knopfli 1990) can be promoted as an integral part of tourism economy. This is important because the observation of handicraft artisans and the buying of souvenirs or art objects draw tourist interest and spending. The government's reserve policy creates parks and reserves that ignore their human neighbours. These, according to evidence from this study are doomed to fail.

SWOT Analysis

The Kimbi Game Reserve within the Nyos Volcanic District is rich in natural and cultural treasures. The strengths (S), weaknesses (W), opportunities (O) and threats (T) of ecotourism resources were identified:

- **Strengths and Advantages**
  - The area is rich in wildlife (plants and animals) and natural landscapes of touristic and scientific importance.
  - It possesses rich cultural values, handicraft production, traditional houses, settlements, festivals, rites and food.
  - Local people are welcoming and have a positive attitude to tourism.
  - Annual festivals and dances are organized in chiefdoms (palaces).
  - Life is traditional and tranquil in a typical African setting.
  - The area has an ecological potential to increase the destinations and experiences for worldwide tourism.

- **Weaknesses or disadvantages**
  - The area is far from the provincial capital city (Bamenda) and other urban centres.
  - There are clearly no tourist infrastructure and service facilities.
  - The reserve is grossly under-staffed with no staff educated for tourism.

- The area is enclaved in difficult topography and remote from urban centres. There will be high investment costs.
- Tourism in the region has no advertisement and marketing activities and agencies.
- Local people and tourists are insensitive to environmental issues.
- No master management and development plans to take advantage of the opportunities offered by ecotourism.
- Lack of tourism marketing and promotion agencies.

- **Opportunities**
    - Integrating rural livelihoods in conservation projects for local employment and poverty alleviation.
    - Promoting, sustaining and reviving a disappearing cultural heritage.
    - Protecting and sustaining the rich biodiversity.
    - Promoting the participation of local people in biodiversity protection.
    - Promoting tourism and biodiversity conservation by linking culture, the environment and livelihoods.

- **Threats**
    - Total absence of infrastructure and waste disposal systems.
    - Increasing human pressure on fauna, flora and natural landscapes.
    - Erosion of the cultural values through production and commercialization to suit the taste of tourists and through the adoption of foreign values.
    - Poorly structured and unplanned village settlements and houses that can be developed as cultural villages.
    - Underdeveloped environmental consciousness and threat of pollution.
    - Risks from catastrophic landslides and gas eruptions from crater lakes.

Conclusion

Tropical countries like Cameroon, beleaguered as they are, have established parks and reserves as a way of saving biodiversity. Take a chunk of forest, the reasoning goes, make it a park or reserve, and charge tourists to visit. But parks that ignore their human neighbours and parks that are poorly protected are doomed to fail. The Kimbi Game Reserve is like most "paper parks" which are littered in the tropics that do not more than protect ravaged natural landscapes and biodiversity. Local inhabitants do not benefit from such national parks in economic terms. The handicrafts common in the past are rarely seen and economic benefits could be gained by reviving traditional handicrafts. The area offers several strengths and opportunities for ecotourism. Careful planning and investments are required to overcome the weaknesses and threats facing its development. In planning most attention should focus on the needs of local people, that is, integrating their culture, livelihoods and environmental awareness in tourism development. In this process local people, government organizations, universities, research institutions and society as a whole must interact sufficiently in order to develop components of interest to ecotourism, ecocultural tourism and scientific tourism.

References

Alpert, P. (1993) Conserving biodiversity in Cameroon. Ambio No. 22, p. 33 – 107

Barkin, D. (1996) Ecotourism A tool for sustainable development. http://www.planeta.com/planeta/96/0596

Chambers, R. (1992) Sustainable livelihoods: The poor's reconciliation of environment and development. In P. Ekins and M. Max-Neff (eds.) *Real-life economics: Understanding wealth creation.* London. Routeledge, p. 214 – 230.

Chifos, C. (2006) Culture – environment and livelihood; potential for crafting sustainable communities in Chiang Mai. *Int. J. Environment and Sustainable Development,* Vol. 5, No. 3. p. 315-332.

Daniel, L.; Manning, R. and Krymkowski, D. (2005) Relationship between visitor-based standards of quality and existing conditions in parks and outdoor recreation. *Leisure Science*, vol. 27, p. 157-173.

Gengiz, T. (2007) Tourism, an ecological approach in protected areas: Keragol-Sahara National Park, Turkey. *Int. J. of Sustainable Development and World Ecology,* vol. 14, No.3, p. 260-267

Gregory, T. (2005) Conflict between global and local land use values in Larvia's Gauja National Park. *Landscape Research,* vol. 30, p. 415-430

Hawkins, P and Brunt, M. (1965) Soils and ecology of West Cameroon, FAO Rome, Report No 2083.

Jurgens, C. R. (1993) Strategic planning for sustainable rural development. *Landscape and Urban Planning,* vol. 27, p.253-258

Knopfli, H. (1990) Crafts and technologies: Some traditional craftsmen of the Western Grassfields of Cameroon. Part 2: Woodcarvers and blacksmiths. Basel Mission, Basel. 125p.

Kwanga, M.J. (2006) Wildlife management: Case Study of Kimbi Game Reserve. Long Essay, Dept. of Geography, University of Yaounde I. 38p.

Leenhartdt, O.; Menard, J. and Temdjim, R. (1990) Rapport sur l'inventaire des lacs maar au Cameroun. Mission Francaise de Coopération à Yaounde. (Manuscript).

Lele, S. (1991) Sustainable development: a critical review. *World Development,* vol. 19, No.6 p. 607-621

McDonought, W. and Braungart, M. (2002) Cradle to cradle: Remaking the way we make things. New York, North Point Press.

Macleod, H. (1986) The conservation of Oku Mountain forest. ICBP Cambridge, Project Report, 90p.

Ndenecho, E. N. (2005) Conserving biodiversity in Africa: Wildlife management in Cameroon. *Loyola Journal of Social Sciences,* vol. No.2, p. 209-228

Ndenecho, E.N. (2007) Population dynamics, rural livelihoods and forest protection projects in sub-Saharan Africa: experiences from Santa, Cameroon, *Int. J. of Sustainable Development and World Ecology*, vol. 14, No. 3, p. 250-259

Neefjes, K. (2000) Environment and livelihoods: strategies for sustainability. London, Oxfam Publishing.

Ngwabuh, B. A. (2002 Annual report of the Kimbi Game Reserve. MINEF; NW. Delegation, Bamenda.

Poiani, K.; Baumgartner, J. Buttnick, S; Gren, S; Hopkins, E; Ivy, G.; Seaton, K.; and Sutter, R. (1998) A scale-independent site conservation planning framework in nature conservation. *Landscape and Urban Planning*, vol. 43, p. 143-156

Redchift, M. (2000) Sustainablity: life chances and livelihoods. London, Routledge.

Sanderson, E.; Redford, K. and Veddez, A. (2002) A conceptual model for conservation planning based on landscape species requirements. *Landscape and Urban Planning;* vol. 38, p. 41-56.

Stuart, S. N. (1986) Conservation of Cameroon Mountain forest. ICBP Cambridge. Project Report.

Chapter Eight

Superficial Deposits and Ground Water Resource Development in the Upper Nun River Valley, Cameroon

Summary
Access to regular, sufficient, clean drinking water and adequate sanitation is one of humanity's basic requirements. In developing countries a significant proportion of diseases and deaths can be attributed to water-related causes. Rural people have little access to this basic resource. The paper uses a combination of field observation, examination, informal interviews and secondary data to assess the influence which geology, superficial deposits and topography exert on ground water levels. It maps the ground water levels on various land facets and correlates these influences with the topography and geology. The study also appraises the development and management of the rural water supply schemes in traditional societies. It concludes that the development of ground water for domestic purposes is not very feasible in the granitic rocks and colluvial materials. The potential is high in alluvial and alluvio-colluvial materials. Finally, it identifies the constraints and institutional deficiencies in the operation of current rural water schemes and recommends that the most feasible option is the development of hand-dug, user-operated and managed wells for groups of hamlets or farm families.

Introduction
Generally, water resource problems often originate from the fact that the hydrological set up of a region does not usually fall into mankind's scheme of spatial distribution, time and quality requirements. So the water problems for various areas have either been too much water as manifested by floods and poor drainage, too little water as indicated by dried up sources during the dry seasons or too little water because of the high degree of pollution of surface and underground water which makes it unsuitable for human consumption. Water schemes in most Grassfield villages of the

North West Region of Cameroon are thus a response to the protracted problem of water scarcity particularly during the dry season. Hence, the development of potable water resources has today become an integral component of the rural economic development by which progress has come to be measured. As village and urban water supplies are generally inadequate and sometimes very poorly distributed (Zimmermann, 1994; Ndenecho, 2003; Acho-chi, 1938), there is a need to search for alternative water resource development options. This is particularly important in rural areas where existing schemes suffer from high construction costs and institutional deficiencies in operation. Rural areas are also seldom the focus of government policy makers (Ndenecho, 2007)

The paper assesses the role of geology and superficial deposits on ground water yields and the implications on the development of rural water supply systems in qualitative terms. Most improved rural water supply projects have not helped the poor. They are unable to fund or contribute meaningfully to current expensive modern systems and the benefits in the form of adequate, regular and good water quality are generally not accessible to all community members. The paper therefore also seeks to identify new water resource development options which can be appropriate and sustainable.

The Study Area

The Upper Nun River valley is an intermontane basin in the Bamenda Highlands of the North West Province of Cameroon (Fig. 1). This valley is a vast plain with an average elevation of 1128 m above seal level. It is bounded to the north, west and east by an escarpment and to the south by Mount Nkogam. Morphologically it is an alluvial plain with varying degrees of drainage interrupted by rounded remnants of small granitic hillocks of 915m to 1120m surface of basement rock. The Nun River flood plain has extensive swamps and seems to have been formed in semi-lacustrine conditions. With a slope gradient of about 1:300 which is only interrupted by the basement complex, the hillocks have been partially buried by alluviums. These protrude as 8 to 16 km long, rounded, finger-like features with crests some 15 to 300 m above the alluvial surface (Fig. 2).

The climate is characterized by a dry season lasting from mid-November to mid-March, and a corresponding rainy season from April to October. Table 1 presents climatic data for Ndop plain. Hawkins and Brunt (1965) described the climate as "very hot and sunny" with mean annual maximum temperatures ranging from 27^{0}C to 33^{0}C and mean minimum temperatures ranging from 7^{0}C to 15^{0}C. The average rainfall ranges between 1270mm to 1778mm.

Babanki-Tungo at an elevation of 1300 to 1500m in the escarpment zone has lower rates of evaporation (1438.4mm / year) than Bambalang which is situated in the centre of the plain (1742.4mm / year). Evaporation rates during the dry season are high. Surface water sources and springs in granitic rocks, and colluvial zones quickly dry out during this season.

Most villages depend on these surface sources which either dry out or are of poor water quality. The hydrological characteristics of these streams have been described by SEDA (1983). The Monkie Stream upstream of Bamunka village has a basin surface area of 226sq. km and an average discharge of 21.9m^3/s. Tatum stream in Balikumbat has a surface area of 21.8sq. km. It has an average discharge of 11.2m^3/s. Tembou stream upstream of Balikumbat has a surface area of 62sq. km with an average discharge of 45m^3/s. Other small streams include Chanke in Bangouren, Monoun in Bangolan, Sefou in Balikumbat and Meyeh in Babungo. Traditionally, rural water requirements are obtained from these streams, springs, ponds in flood plain swamps and harvested rain-water. Thirteen villages in the area over the years have invested community labour and financial resources to avail themselves of regular, potable water. The average population density is 76 inhabitants/sq. km.

Fig. 1: Location of the Upper Nun River Valley and Study Sites Evaporation has been recorded in Bambalang and Babanki-Tungo (Seda, 1983). The data is presented in Table 1.

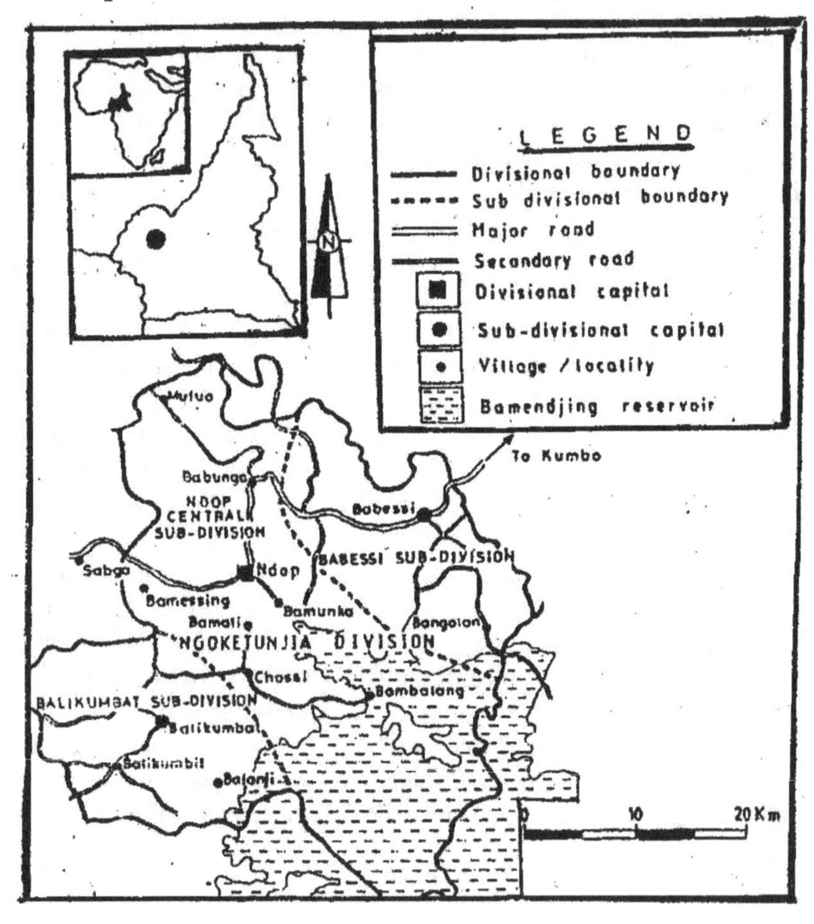

From the foregoing presentation, the study area possesses abundant water resources which together with rainy season flooding of the plain have been described as a "water empire" (Lambi, 1999; Lambi, 2001). Unfortunately, potable water resources are not preponderant. Villages with no access to gravity water schemes and pump-assisted groundwater schemes depend on wells and intermittent springs in *Raphia vinifera* swamp –forest within the finger-like granitic hillocks Fig. 2 & 3). Village water supply schemes suffer from management problems, lack of spare parts and lack of technical assistance. Ndenecho (2007) recorded 14 drinking water supply services in use and 6 broken down rural water supply schemes. The National Water Corporation (SNEC) which caters for urban water requirements operates in Ndop town (regional capital). Ndop has water storage problems (MINPAT/UNDP, 1999). Despite its production capacity of $600m^3$/day, the storage reservoir into which water is pumped for eventual distribution has only a capacity of $100m^3$/day. Ndop town therefore, suffers from acute water shortages. The rural and urban population, therefore, need alternative water development and management options.

Methods and Data Sources

In order to assess the influence which surface relief or topography and geology exert on the levels at which ground water will be found, the study area was mapped into its major land systems and land facets (Dent and Young 1981). Geological and topographical maps and aerial photographs were used to realize a reconnaissance survey mapping intended to provide a broad view of the ground water resources of the area. Various field sites and existing wells on landforms (land facets or land units) on various land systems were identified in order to establish a hydrological survey of the area. The depths of water per existing well level were measured and then related to the geology and nature of superficial deposits. Brief profile descriptions of randomly selected sites on the various land facets were made during the sinking of boreholes or and observing the samples. Some parameters involved included the parent material, slope, runoff and drainage, colour, hand texture, permeability, structure, internal drainage and vegetation. In this way, the general pattern of ground water levels and seasonal availability in the area was established. Informal interviews of stakeholders of rural water supply schemes yielded data that enabled an appraisal of the various water sources. The above data were supplemented with field observations and secondary data sources. It must be noted that this hydrogeological study is a reconnaissance survey. It presents only a broad view of the ground water resources of the area.

Table 1a: Climatic data for some stations in the upper Nun Valley

Parameters For Ndop	J	F	M	A	M	J	J	A	S	O	N	D	Annual Total/ average
Relative Humidity (%)	57	52	66	82	87	96	93	94	93	90	81	65	79
Max Temperature (°C)	29	29.8	28.9	27.5	26.4	25.2	24.5	24.5	24.8	25.5	26.9	28.2	26.7
Min temperature (°C)	15.0	15.6	16.8	16.9	16.6	15.9	16.1	15.7	18.8	18.8	15.5	15.0	18.5
Average temperature (°C)	22	22.7	22.6	22.2	21.5	20.6	20.3	20.1	20.3	20.7	21.0	21.6	21.3
Sunshine hours hours/days	8.3	8.75	6.91	6.57	6.83	5.83	4.07	3.93	4.44	5.87	7.78	8.63	6.47

Table 1b: Evaporation data for the Nun River Valley (mm / month)

Station	J	F	M	A	M	J	J	A	S	O	N	D	Total
Babanki-Tungo	131.7	162.7	141	131.5	136.5	119	96	93	96	114.7	103	114.7	1438.4
Bambalang	173.6	199.7	190.6	145.5	148.8	138	102.3	96.1	100	130.0	144	173.6	1742.4

Results and Discussions

Figure 2 presents the geology of the study area while Fig. 3 shows the topomorphic units. The geology and topographic sites were assessed in terms of ground water resource potentials. This section gives a correlation between the underlying geology, the topographic surface forms (Landforms or topomorphic units) and their ground water potentials. The hydro-geological characteristics are:

Organic materials

These are raw peats in sedge and grass swamps. These were observed in the Nun River flood plain in almost flat topography. The parent material is alluvium in the upper edge of the flood plain. Runoff is very slow and the area experiences seasonal flooding. The water table is 90cm from the surface. The profile is dark-brown to grey-brown with a silty loam top soil, which changes to silty clay at depths above 47cm. It is structureless and massive with occasional distinct vertical cracks. Small mica flakes are common throughout the profile. Water definitely appears to drain along cracks. Spring sources occur at the base of thalwegs. During the wet season these are inundated. There are open pools of water but these suffer from pollution and nutrient enrichment from rice, food crop fields and the watering of cattle. These are the main features of flood plain and meander belts. Alluvial deposits are the main superficial deposits in these topographic sites.

Alluvio-colluvial deposits

They occur in broad plains in narrow valley bottoms between the granitic hillocks. These are gallery swamp forests with spring sources. In these sites mounds of micro-relief accumulate around the breathing roots of larger trees, raphia palms and date palms. During the rainy season, pools of stagnant water may accumulate in the depressions in micro-relief. Most of this water drains during the dry season, but the soil remains waterlogged. (Ndenecho, 2007). The texture increases quickly from a silty clay loam at the surface to a very sticky and plastic clay at greater depths. During the dry season the water table is less than 1.5m from the surface. Some spring sources may dry out by the end of the dry season.

Colluvial deposits

Colluvial deposits form long concave foot slopes and broad plains (Fig. 2 & 3) along the foot of the escarpment and valleys. The area is gently sloping and the parent material is either granitic or colluvium as in Ndop or lava colluvium as in Babungo. Mid slopes were investigated. The area is well drained with rapid runoff. Granitic superficial deposits present a massive structure, with weak sandy loam in the topsoil (coarse sand fraction), coarse sandy loams in the subsoil and sandy clay loam (gritty) at deeper layers. The lava colluviums on the other hand are massive with vertical cracks in the top layers. The structure becomes blocky at deeper layers. These are highly permeable. The subsoil structure closely resembles the structure of weathered basaltic boulders seen near Jakiri. Granitic colluvium experiences gullying to a depth of 10m. The water table is present at depths of 22 to 25m.

Granitic basement

These are planation surfaces or hillocks. This formation of coarse granite is found just beyond Bamali palace. It is undulating but near rocky hills and constitutes the basement complex. The area is well-drained and runoff is rapid. The structure is massive with traces of rock-like material, large mica particles and stone-lines composed of quartz particles. The texture is sandy loam in the topsoil, sandy clay loam in the subsoil and sandy clay at deeper layers. Permeability is high and internal drainage rapid. The potential for the development of ground water in this land unit is low and not very feasible for hand-dug wells because ground water exists at depths greater than 25m below the surface.

Mineral alluvium

These are found in mineral swamp forests. The parent material is alluvium on gentle slopes. This unit also has micro relief mounds around trees. Runoff is ponded and the drainage is swamp. Internal drainage is slow and fine mica is found throughout the profile. The entire profile is grey with a sticky plastic consistency. The land unit is liable to season flooding and the ponded water to nutrient and sediment pollution. The water table is less than 1.5 metre from the surface. Water availability in this land unit is seasonal.

Fig. 2: Geology, Landform and Hydrogeological Characteristics of Various Land Facets

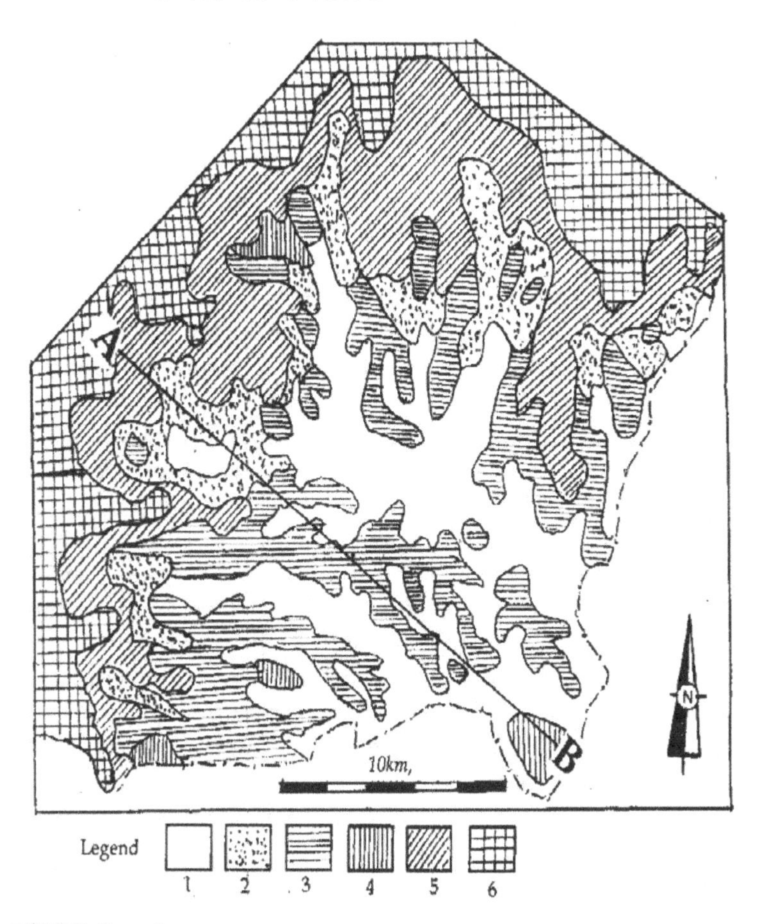

Legend 1 2 3 4 5 6

LEGEND: Figure 2

Map code	Geology	Landform	Groundwater Potential Depth to water table
1	Lower alluvium	Plain	< 3 metres
2	Upper alluvium	Colluvial foothills	22 - 25 metres
3	Acid plutonic rocks: Granite + Seynite	Hillocks	22 - 25 metres
4	Basalt	Hills/Mountains	Not very feasible
5	Gneiss/Micaschist	Escarpment	* Not very feasible
6	Trachyte/Rhyolite	Lava Plateau	Not very feasible

Cross section: A - B: See Figure 3

** Locations operating gravity water schemes*

The surface relief or topography generally exerts an important influence on the levels at which ground water would be found (Ayoade, 1988). Since the configuration of the water table follows the nature of the initial topography of the landscape, the ground water table is closer to the elevated interfluves. So the depth of the water table increases with increasing distance away from the stream channel or the outcrops of the springs (Ruxton and Bery, 1957; Ruxton and Berry, 1961). Consequently, the depth of the water table increases with increasing distance away from the stream channels. A borehole revealed the depth of the weathering profile in the underlying basement granitic rock. While many areas of Babungo are blanketed by volcanic lava of basaltic composition, this residual hill range is one of the outcrops of the ancient basement complex. Boreholes need to be as deep as 25m in order to hit the water table in the following locations in Babungo: Magnolia Estate, Government Health Centre and Government Secondary School. In Ndop around King's Heritage Hotel a borehole to a depth of 11.5m struck the water table but this was polluted by local sewage systems. Site selection and depth of well are important considerations to make. Highly jointed rocks are conducive for underground water resources. The rocks are all strongly jointed granites. Lambi (1990) established that the basement rocks and their overlying volcanics are criss-crossed by a joint network, and this structural characteristic enhances the flow of underground water resources. In sandy particles, and fractured or fissured granites, the rate of underground water transmission could be as much as 27 million litres in 70 hours (Mabbutt, 1952; Mabbutt, 1961)

Experiences in Babungo village by an international non governmental organization (Plan International) present the following lessons.

- If the resistivity sounding method established that the permanent water table was located at a given depth, this Resistivity Sounding Method or the Schlumberger Array Method can be used successfully to determine the depth of the water table below the surface. Usually, the engineers and geologists do the resistivity sounding test over a radius of 50m round the chosen area in order to establish which points or areas have the water table closer to the surface. This method

Fig. 3A: Cross Section of Ndop Showing the Distribution of Floristic Communities per Edaphic Condition and Topographic Site

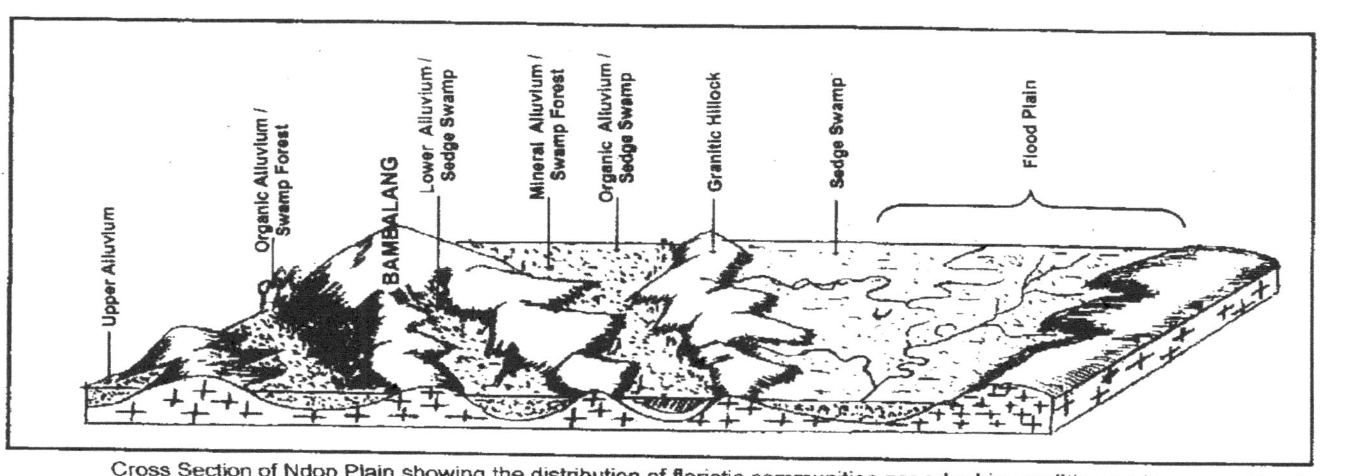

Cross Section of Ndop Plain showing the distribution of floristic communities per edaphic condition and topographic site

Fig. 3B: Cross-Section A-B Showing the Various Land Facets and Associated Superficial Deposits

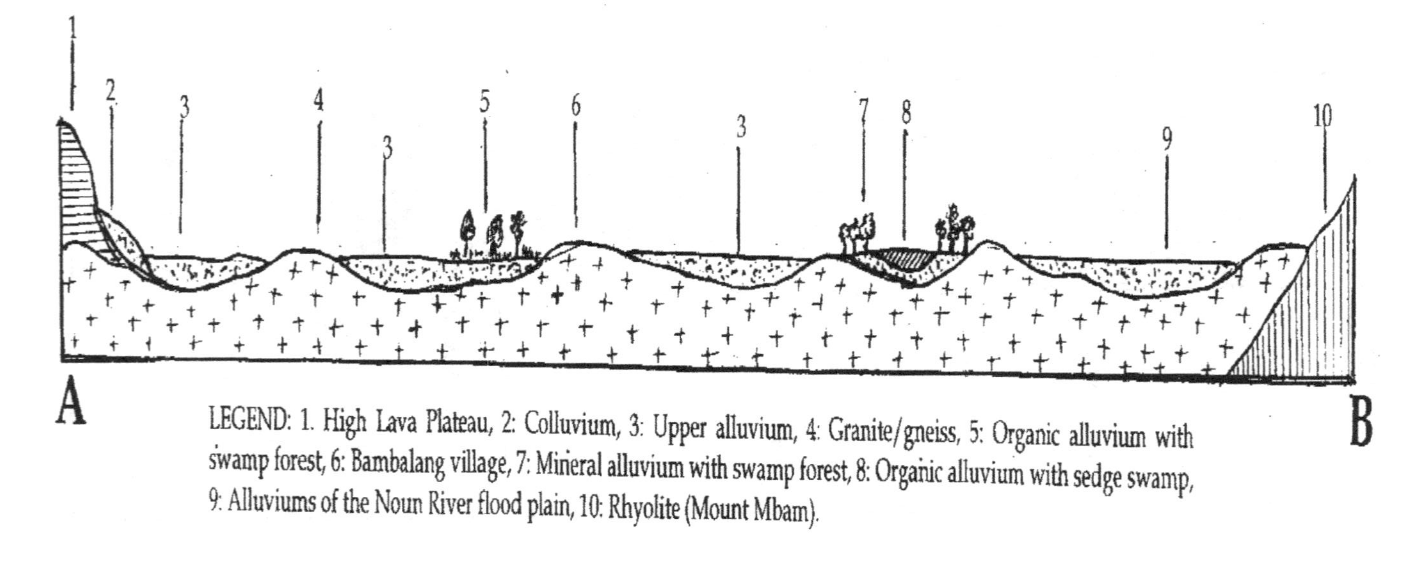

LEGEND: 1. High Lava Plateau, 2: Colluvium, 3: Upper alluvium, 4: Granite/gneiss, 5: Organic alluvium with swamp forest, 6: Bambalang village, 7: Mineral alluvium with swamp forest, 8: Organic alluvium with sedge swamp, 9: Alluviums of the Noun River flood plain, 10: Rhyolite (Mount Mbam).

is particularly useful in areas with granitic rocks and collovial materials. On such sites the water table is generally more than 25m below the surface.

- There are no industries or other sources of pollutants whatsoever around the area which might be responsible for the pollution of the underground water resources.
- The establishment of these boreholes requires that they be located at least some 16m away from any soak away. At the various sites on this hillock, the nearest soak away is at least 30m away.

The development of rural water resources in the Upper Nun River Valley presents the following options:

- Development of surface water sources based on gravity and sand filter system in the colluvial belts (Zimmermann, 1996).
- The combination of gravity system with assisted water pumps to lift water to higher locations in colluvial zones and granitic hillocks
- The combination of wells and pump systems. Boreholes are equipped with electrical pumps to lift water to reserve tanks and reservoirs for redistribution.
- The use of boreholes furnished with pulley systems for groups of farm families. These are simple hand-dug wells. These are cost effective and appropriate.

Experiences over the past decades have shown that the third option was adopted by a Canadian-assisted project known as SCAN Water in Bambalang village. The system was not sustainable due to the inability by local communities to procure diesel and oil, spare parts and technical expertise required for maintenance. In Babungo village a community water supply scheme based on gravity left parts of village frustrated as water could not reach higher elevations. The second option was therefore adopted. This option suffers from contamination and siltation in streams. This complicates water use and requires comprehensive treatment. The care takers are not able to maintain the sand filter tanks and everywhere additional gravel filters have to be built. The gravel filters are very effective but to avoid further problems in future, farming practices in watersheds must concentrate on soil conservation activities which control runoff and gully erosion. High construction costs and institutional deficiencies in operation of gravity systems and pump assisted systems cause further problems. The above problems stem from too few and inadequately trained personnel to plan, implement, manage and monitor comprehensive rural water development programmes.

Table 2: Summary of superficial deposits and ground water potentials in the Upper Nun River Valley

S/N	Superficial Deposits	Vegetation	Observations
1.	Organic alluvium	Gallery swamp forest/fern and sedge swamp	Depth of water table = 90 cm
2.	Mineral alluvium	Gallery swamp forest/tall grass and sedge swamp	Depth of water table = less than 1.5 metres
3.	Alluvio-colluvial deposits	Forest galleries derived from moist evergreen forest	Depth of water table = less than 1.5 metres (seasonal springs)
4.	Colluvial deposits	Wooded savannah derived from evergreen forest	Depth of water table: 22-25 metres (seasonal springs)
5.	Granitic basement	Derived wooded savannahs from evergreen forest	Very low (seasonal springs)

The most appropriate technology required in the plain land units is the development of user-operated boreholes. Plan International in response to these constraints to the development of rural water supply schemes in the area initiated hydro-geological surveys for villages in the granitic hillocks, that is, Bambalang, Babungo, Bamali and Bamunka among others. The project seeks to develop user-operated boreholes for groups of hamlets. In the alluvial and alluvio-colluvial land units the drilling of boreholes is very feasible and cheap but rural people must be made aware of the site selection criteria, sanitation and management. Due to the high flood risk, these considerations became very important. Moreover, shallow and poorly managed wells can be breeding grounds for malaria and typhoid fever vectors which are common in the area.

Conclusion

In Cameroon, more attention has been paid by government to the provision of potable water to urban centres. Rural water supply development has been the concern of local and international non-governmental organizations and the Community Development Department in partnership with village communities or stakeholders. The development and management of rural water supply schemes has been plagued by high construction costs, and institutional deficiencies in the operation of gravity systems and pump-assisted water delivery systems. The problems stem from the few and inadequately trained rural development personnel that can plan, implement, manage and monitor comprehensive rural water development programememes. These considerations are a *raison d'être* for the development of user-operated community boreholes where ground water resources are both available and unpolluted. Traditionally, these have proved to be sustainable in many villages in the Upper Nun River Valley of Cameroon.

References

Acho-chi, C. (1983) Spatio-temporal analysis of rural water schemes in Cameroon Grassfields. Unpublished Ph.D. Thesis, University of Ibandan. p. 40-43

Ayoade, J.O. (1988) Tropical Hydrology and Water Resources. Macmillan Publishers Ltd., London

Dent, D. and Young, A. (1981) Soil surveys and land evaluation. George Allen and Unwin, London. p. 104-106.

Hawkins, P. and Brunt, M. (1965) Soils and ecology of West Cameroon. FAO Rome.

Lambi, C.M. (1990): *Geological Influences on Landform Development in the Bamenda Highlands of Cameroon.* Unpublished Ph.D Thesis, University of Salford, Lancashire, England

Lambi, C.M. (1999): *The Bamendjim Dam of the Upper Noun Valley of Cameroon. No Human Paradise* In: J. Dunlop and W. Roy (eds) Reader in Environmental Education, University of Buea / University of Strathclde, Glasgow.

Lambi, C. M. (2001): Environmental Constraints and Indigenous Agricultural Intensification in Ndop Plain (Upper Noun Valley of Cameroon) In: Readings in Geography. C. M. Lambi and Eze B. E. (eds). Unique Printers, Bamenda.

Mabbutt, J.A. (1952): A Study of Granitic Relief from South West Africa. Geological Magazine, Vol. 89, p 87-96

Mabbutt, J.A. (1961 b): "Basal Surface or Weathering Front" Proceedings of the Geological Association, London, Volume 72, p 3 57-8

MINPAT/UNDP (1999) Regional Socioeconomic study on Cameroon: The North West Province. MINPAT/UNDP Project Yaounde.

Ndenecho, E. N. (2003) A landscape ecological analysis of the Bamenda Highlands. Unpublished Ph.D. Thesis, University of Buea.

Ndenecho, E. N. (2007) Biogeographical and ethnobotanical analysis of the Raphia palm in the West Cameroon Highlands. *Journal of the Cameroon Academy of Sciences,* vol. 7. No. 1, p. 21-32.

Ruxton, B.P., & Berry, L. (1957): *The Weathering of Granite and Associated Erosional Features in Hong Kong.* Bull. Geol. Soc. America, Vol. 68, p. 63-92

Ruxton, B.P., & Berry, L. (1961b): Weathering Profiles and geomorphic position and Granite in Two Tropical Regions. *Révue de Géomorphologie Dynamique*, Volume 12, p. 16-31

SEDA (1983) Etude d'intensification du sous projet dévéloppement rural intègre perimètre Balikumbat-Bambalang. ORSTOM/SEDA, Yaounde. p. 13-16

Zimmermann, T. (1994) An introduction to watershed management for intake areas of rural water supplies in Cameroon. *Proceedings of Agroforestry Harmonisation Workshop,* April 1984, Regional College of Agriculture Bambili, p. 20-27

Zimmermann, T. (1996) Watershed resource management in the Western Highlands. Helvetas-Cameroon, Bamenda, 67p.

Chapter Nine

The Babungo Pipe-borne Water Project: A Community Self-reliant Development Scheme in the North West Province of Cameroon

Summary

In 1980, the inauguration of the Babungo water project showed the impact of the distribution of drinking water in a rural area. The combined efforts of the government and the village community and its association of external elites ensured the success of this water project; this example of self-help development has undoubtedly not only transformed and ameliorated the living standards of the rural population, but also reduced the daily burden of women and children fetching water for households. The completion of this pipe-borne water project in this small rural environment in the Bamenda highland region of Cameroon was an example which served as an inspiration for other village communities in the Upper Nun valley which also set up their own water projects.

Introduction

Good drinking water is a scarce natural resource the world over. According to estimates by the Joint Monitoring Programme of the WHO and UNICEF, 900 million people worldwide will not have reliable access to safe drinking water in 2015; and twice that number will lack adequate sanitation. These figures, moreover, are optimistic in so far as they assume that the infrastructure in place today will remain fully operational in the long term. However, Funke-Bartz (2008) warns that this should not to be taken for granted because the galloping population growth today does not seem to be marched by an equivalent provision of the necessary social infrastructure. Moreover, the caprices of the on-going climate change are all clear evidences at our disposal that potable water will continue to remain a scarce resource except governments, municipalities and the rural communities take upon themselves this water challenge, which is a looming problem for humankind.

The Babungo gravity pipe-borne water project was launched way back in 1980 by the Village Development Committee in an effort to provide clean, safe and readily accessible drinking water for its rural population, which at that time, depended on distant water sources with water supplies of doubtful quality. This innovation was, indeed, a welcome relief as it saved the population from the scourge of water-borne diseases.

Unlike other regions of Cameroon where galloping population growth and the establishment of industries have sometimes called for increased water supplies, or dry Sahelian margins where inadequate, sporadic and unreliable rainfall has necessitated the building of water projects to satisfy basic drinking needs, and the irrigation of thirsty agricultural fields, the Babungo water project was not the result of harsh ecological environment in which surface water is scarce. So in spite of the Babungo abundant water supplies,, most people had no access to clean and proximal water resources.

Babungo in the Upper Nun Valley, is situated on one of the well-watered southern slopes of the Bamenda Volcanic Highland region of Cameroon (Fig. 1). Evidently, the problem here that motivated the pipe-borne water project was not the scarcity of water *per se* as a basic essential for human consumption. Rather the scheme was launched in 1980 by the Babungo Development committee primarily with the view to providing clean, safe, and accessible drinking water for its 20,000 inhabitants, and for desirous neighbouring villages should the water and gravitational force be in excess of the village requirements. More than 95% of the population had hitherto depended on occasionally polluted streams for its water supply. Furthermore, it was hoped that once this giant welfare scheme was completed, it would serve as an impetus or spring board for further rural self-help development programmes in the village.

It was in pursuance of this goal that a network of hundreds of pipelines covering several kilometres was implanted spreading the length and breadth of the village. When the implications of this phenomenon are examined, for the 20,000 inhabitants, the freedom from the scourge of water-borne diseases which are well-known occurrences in this part of the Upper Nun Valley stands out as the most significant benefit from the project. Proper and regular supplies of drinking water are, indeed, some of the most important aspects of the well-being of people. Drinking water is a basic necessity of life (The Courier, 1980).

Fig. 1: The Hydrographic and Settlement Network of Babungo, Ngoketunjia

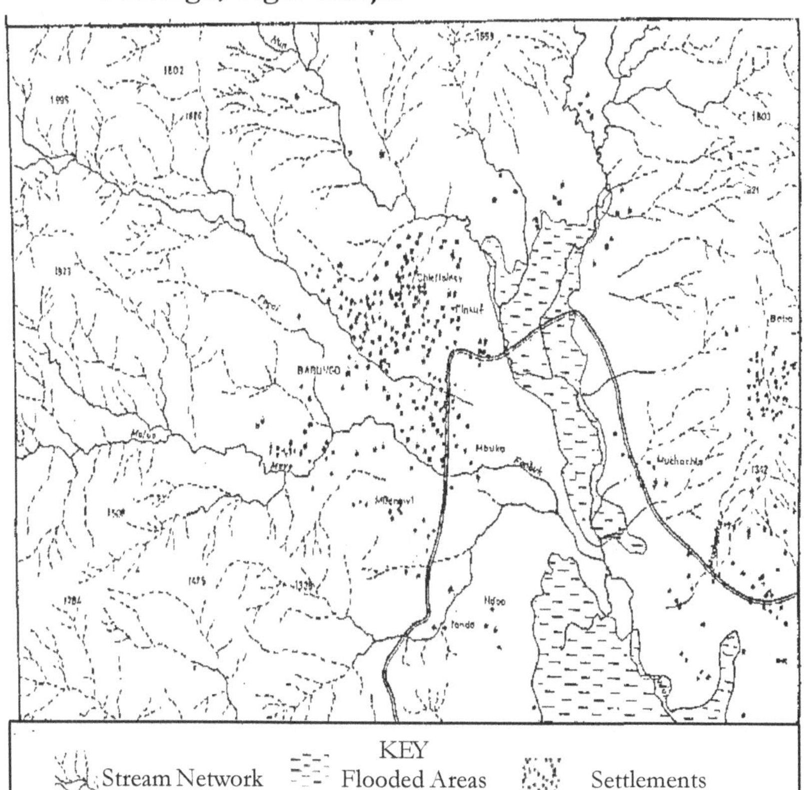

This paper examines the location of settlements in relation to the previous sources of water supply, the strategy and technology of the gravity water scheme, the funding and significance of this rural development project in this part of the volcanic highland area of the North West Region (Fig 2).

It also shows the role that misappropriation of community funds can play in the stagnation of a viable and valuable development project.

Fig .2: Location of the Study Area in the Bamenda Highlands of Cameroon

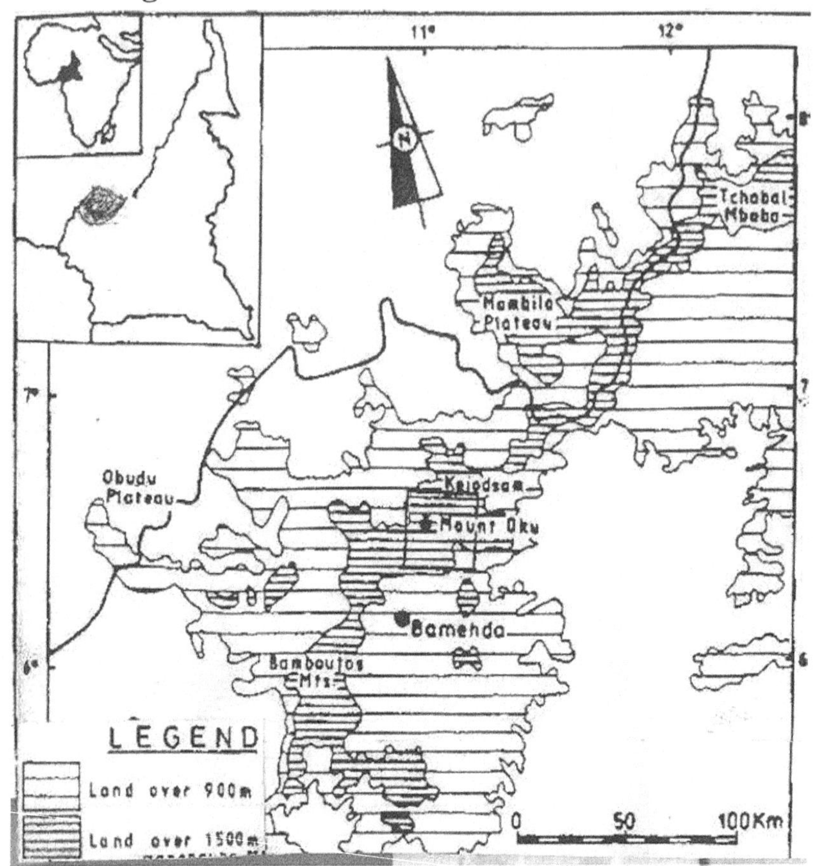

Settlement and Sources of Water Supply

While the location of the settlements may be determined by the availability of water in some parts of Cameroon, it seems a living anachronism that the Babungo administrative wards have grown and expanded without recourse to pure drinkable sources of water. But Babungo is not alone. Like many other rural areas, the main problem has remained the availability of water irrespective of its quality and distance to such sources of water supply. Table 1 gives the list of quarters, the mean distances between the different sources of available water supply and the settlement concentrations in Babungo.

Prior to the development of the Babungo gravity water project, the village sources of water consisted of streams, forest and hillside springs, raphia bushes, swamps, intermittent or periodic streams and wells. Not only are most of these sources highly polluted during the rainy season, several of them are indeed, distally located from the centres of high population concentrations (Fig. 3). While it was obvious that the use of deep, well-protected wells could solve the recurrent problem of drinking water from occasionally polluted sources of water, wells were, however, not an ubiquitous phenomenon in the village. In fact, less than 3% of the population depended on wells for the household water supply.

Although the bulk of the 20,000 people use the other sources of water supply, some areas, however, are favourably located in terms of source and purity of streams. The areas located on the foothill zones of the village where fresh water springs gush out of the volcanic or granitic mountain sides have no problems in terms of quality and distance from source area.. Where the watersheds upstream are not grazing grounds for Fulani cattle, such areas would also be free from pollution. However, field survey shows a very low population density in these favourably located settlements or village administrative wards.

Perhaps, the most suitably located in this respect in terms of nearness (100-300m) and purity are the four upper catchment areas, namely, Toh Finkwi, Toh Saji, Toh Sofi, and Mbenje (Fig. 3). These peripheral areas (Toh in Babungo) have limited or no farmlands. Hence, water pollution by sediments from the farmlands is limited. At worst, occasional and insignificant pollution may come from the watering of cattle upstream.

Three quarters, namely, Veubwi, Mbenje and Loong have been classified, for the purpose of this study, as water empires because they make up part of the permanently flooded Upper Nun Valley, and these quarters also possess streams which make up the tributaries which take their rise from the surrounding volcanic and granitic ranges (Fig. 2). So the availability of water from the above nearby source areas is a foregone conclusion. The mean distances to sources of water lie between 100-150m. The second class of proximal source of water supply are Mbuokang, Ibia, Tavendong and Ibau. These quarters are located on the interfluves of main streams, and the mean distances from the centre (heart) of the quarter to the nearest streams range from 200-300m from Ibau, and 300-500m from Ibia and Mbuokang. Figure 3 shows the distribution of the quarters in relation to the availability of water supply.

Table 1: Mean Distances to Nearest Available Sources of Quarter Water Supply

Quaters	Mean Distances	Sources of Water Supply
Mbungwi	0.75 – 1.5 km	Stream and raphia bush spring
Tondo/ Mbekong	1.0 – 1.75 km	Stream and periodic spring
Finkwi	1.0 –1.5 km	Stream
Finteng	1.0 –1.5 km	Stream and forest spring
Mbuokang	300 –500m	Streams (Interfluves location)
Ibau	200 –300m	Stream
Ibia	300 –500m	Streams (Interfluves location)
Mbenje	100 –150m	Water Empire (streams)
Veubwi	100 –150m	Water Empire (streams)
Tavendong	250 –350m	Stream
Loong	100 –200m	Water Empire (streams)
Toh Finkwi	200 –300m	Stream
Toh Saji	200 –300m	Stream
Toh Sofi	100 –200m	Hillside spring and stream
Ngoleh	300 –400m	Swamps and periodic stream

The bigger the sizes or radii of circles, the farther are the distances from the foci of the quarter to the sources of water supply. On the other hand, the smaller the circles, the nearer are the sources of the water supply to the core of the quarter.

Perhaps, the cases of Finkwi, Finteng, Tondo/Mbekong and Mbungwi could be highlighted to illustrate the water problems of the administrative wards. These four areas experience water problems both in terms of distance and quality. Finkwi and Finteng which constitute the heart of the population core of Babungo (Fig. 3) are some of the most distally placed from available sources of water. These two population hubs depend on streams whose mean distances are approximately 1.0 -1.5km away.

Mbungwi depends on periodic streams and raphia bush spring. The former dry up during the dry season leaving the inhabitants to depend on the raphia bush source or travel 0.75 -1.5km to fetch water from the nearest stream. Tondo and Mbenkong are worse off in terms of water supply. However, a point to emphasize is the fact that these are all periodically contaminated sources of water, and the distal location of settlements from any good drinkable sources of water supply necessitated the establishment of the Babungo gravity water project.

Strategy And Technology

The original strategy of this self-help development project was to provide affordable, good drinking water for the villagers by the end of 1984. Following the application of the village Development Committee in 1980 for the construction of a water scheme, the Technical Staff of the Bamenda Department of Community Development, an organ of the Ministry of Agriculture, carried out feasibility studies during which four springs were seen to have a capacity worthy of providing the village with enough water even throughout the peak of the dry season. And it was estimated that on completion, daily consumption per person per day would be nearly 130 litres. Moreover, this scheme was conceived to operate on gravity control so as to eliminate the use of mechanical or hand pumps as it is the practice in some areas where available catchments are sited on lowlands and are consequently devoid of gravitational force to redistribute the water. This explains why on the High Plateau where the springs formerly gushed out from the volcanic rocks, there are four catchments with a flow rate of 2 litres per second in the dry season, four interruption chambers, and a huge magical main tank which roughly handles 80.000 litres of clean water where purification on treatment with chemicals is not necessary.

It was thus estimated that on completion of the project (Progress Report), it would consist of four spring catchments (all of which will be capable of supplying 2 litres/second in the dry season), 4 interruption chambers, a huge storage tank (capable of holding 80.000 litres), 12 stand pipes, 12 wash basins, 5 fountains, 4 coffee wash places, and the total project will have a distribution network of about 16.000m pipes.

Fig. 3: The Map of Babungo Showing the Mean Distances to the Nearest Available Sources of Water Supply

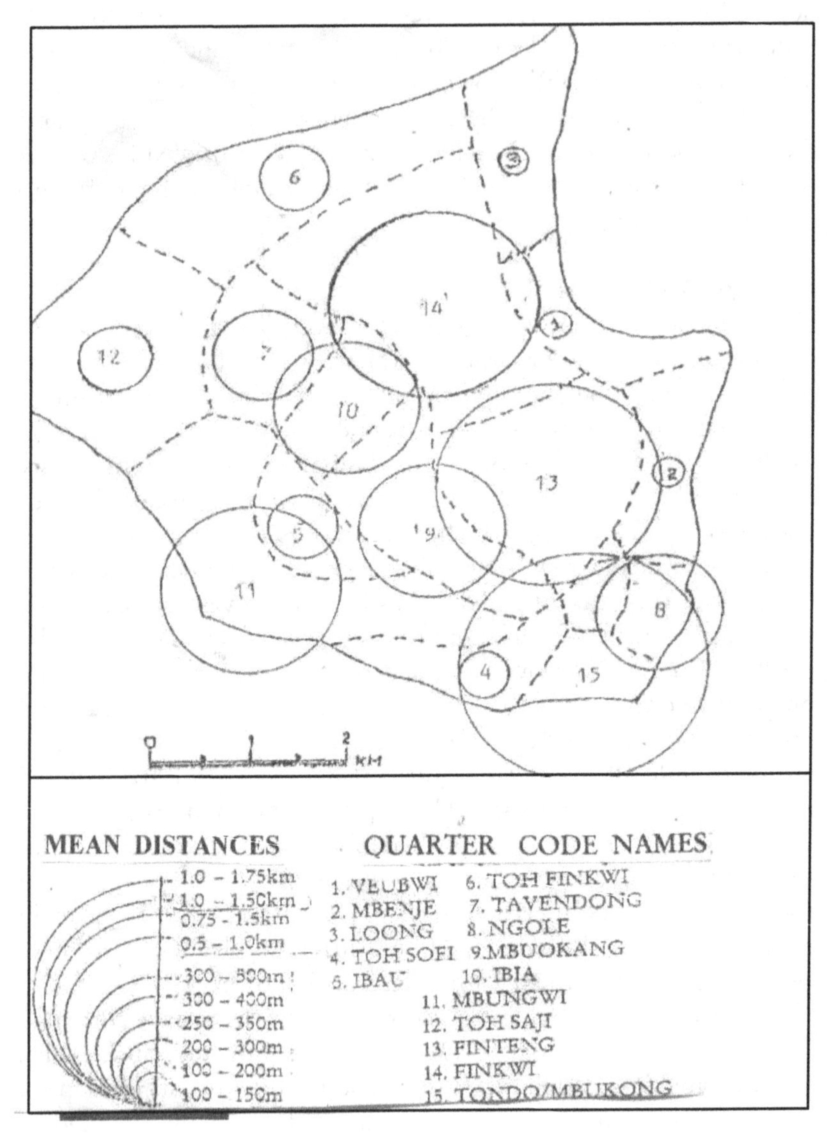

The completion of the above program was to be achieved in two phases. Phase One of the project which covered the 1981/1982 and 1982/1983 years saw the construction of the four catchments, all the interruption chambers, and the huge magical storage tank, as well as the construction of three of each of the stand pipes, wash basins and fountains. 7.000m of pipe line out of the estimated 16.000m were connected from the storage tank through the market to the Palace, thanks to MIDENO and SATA. Also in view of the great pressure with which the water flows down into the tank, safety valves were constructed through which water above a certain height in the storage tank drained away freely. Without this precautionary measure, excess water pressure could break down the storage tank. However, it was estimated that when all the administrative wards of the village are supplied with pipe-borne water on the completion of Phase Two, the safety valves could then be closed because the total consumption by all the quarters should normally leave the water in the storage tank at comfortably low ebb.

After the construction of the four catchment tanks at the sources of the spring, under technical supervision, grass was replanted over the areas to conceal the tanks so that the sites have the grassland scenery and that is much akin to that of the surrounding countryside. The essence of this exercise was to prevent any pollution by cattle since the highland source areas remain the domain of the Bororo cattle kingdom. The daily consumption was estimated at nearly 130 litres per person at the completion of the whole project.

The second phase of the gravity water supply scheme entailed the rounding up of the uncompleted items as specified in the project description above. This phase was to enable the remaining parts of the village to be supplied with water as, hitherto, only those inhabitants who were privileged to be between the catchment region and the main market centre as well as a few other wards, had water supplies. This extension, according to technical estimates, would require a further 17.633.780 franc. These estimates were based on the November 1983 unit prices including 10% increase to make room for inflationary trends. This is discussed under the funding of the gravity water supply.

Statistics show that the citizenry of the village had to make available in cash the sum of 14.250.000 francs while 5.000.000 francs were to be provided in kind – which had to be the provision of manual labour from time to time. Village voluntary labour and other payments in kind contributed enormously to the success of the Babungo Self-Help project. In this way, the project became easily acceptable and consequently financially more affordable to the village low income farmers and petty traders. One fact that emerged from the execution of this project was that rural involvement was a determining factors that off-set financial difficulties.

Table 2: The Financing Plan of the Babungo Gravity Water Supply Project

Participant/ Body	%	Kind	Cash	Total (Fcfa)
Community Communauté)	35	5.000.000	14.250.000	19.250.000
State (Etat)	28	1.600.000	9.400.000	11.000.000
Foreign Aid/ (Aid exterieur)	40	__	22.000.000	22.000.000
Others (Autres)	5	----	2.750.000	2.750.000
			Total	**55.000.000**

The realization of increasing total annual amounts of money collected in the first phase of the project despite the stagnation of yearly levies is an indication of the awareness of the importance of this project by the citizens both resident at home and abroad. There were, however, significant increases from external contributors (migrant citizens in the big urban centres) probably because they were satisfied with the initial phase of the work and probably because of the correct management and financial accountability of the whole project. No one, of course, would have cheerfully continued to replenish funds into mismanaged coffers under the name of rural development. So the big financial subscribers to this project are those citizens resident outside the village.

Table 3: Investment Plan of the Project

Participant/ Body	1981/82	1982/83	1983/84	1984/85	Total (Fcfa)
Community (Communauté)	5.000.000	5.000.000	3.000.000	1.250.000	14.250.000
State (Etat)	2.500.000	2.500.000	2.500.000	1.900.000	9.400.000
Foreign Aid/ (Aid Exterieur)	12.000.000	5.000.000	5.000.000	—	22.000.000
Others (Autres)	1.000.000	1.000.000	750.000	---	2.750.000

On the local front, the villagers accepted this innovation especially when they realized that the migrant contributors were pumping in more money than those of them who were the direct or primary beneficiaries from the water supply scheme. The amounts were raised through annual levies based on the categories of each citizen whether resident at home or abroad. Men and women at home were given smaller levies primarily because they have a low income and partially because they also provide manual labour. The fact that more money continued to pour in from the migrant citizens from the major cities of the Republic such as Douala, Yaounde, Limbe, Buea and Bamenda, only to name a few, showed the growing interest they had in the development project.

Furthermore, the aid from the North West Development Authority – *Mission de Dévélopement de la Province de Nord Ouest*, MIDENO, another organ of the Ministry of Agriculture whose motto is *Excellence in Rural Development* was enormous (Tables 2 & 3). This, as well, provided evidence of government support and interest in these rural development projects whose ultimate goal satisfies one of the numerous aspects of government's development objectives, namely, *Health for all by the year 2000 A.D.* throughout the national territory of Cameroon. This national objective falls in line with the WHO and UNICEF Millennium goal of providing reliable and safe drinking water for considerable large number of people by the year 2015. Moreover, it is reckoned that the main objective of MIDENO is to improve the living standards of the one million people of the North West Province. And one of the ways of achieving this is the provision of good drinking water which is either lacking in some parts of the province or grossly inadequate.

It also undertakes the improvement of village water supply where water is available but of poor quality. This is why the Ministry of Agriculture through MIDENO and the Department of Community Development did put in a sum of 7.283.770 FCFA out of the 18.283.770 for 1981/83 either by direct cash or by material aid (Tables 2&3).

Stagnation of the Project

An examination of Table 2, shows that the last general contributions which were ploughed into the project were in the 1984/85 financial year. But contributions towards the water project did not stop with the 1984/85 year. Standardized levies and generous contributions continued to come in but these amounts were simply siphoned by officials who were supposed to be the custodians and the managing group of the development scheme. The post 1986 period actually coincides with the change of financial management of the project. Some dignitaries either fraudulently become signatories to the account or totally hijacked the development treasury. Consequently, no accounts were given to the contributors any more. The loss of confidence in management reduced the annual inflow of money while levies for rural development literally stopped.

So it becomes clear once the open financial embezzlement syndrome set in, the progress of the water project quickly came to a standstill. Even the stockpile of plastic water pipes which had been bought and transported into the village for use in the further extension of the pipeline network to the other areas which had not been previously served during the first phase laid idle and the projected extension works became an illusion. At best, individuals who wished to draw the pipe-borne water to their premises and houses had to buy their own pipes before paying the connection fee of 30.000 francs which like other monies, ended up in some private pockets rather than the central funds of the project. These poor financial practices and gross mismanagement of public funds ushered the unmistakable halt in the development project from 1988.

The quasi-collapse of the development scheme remained a clear example of a crisis of confidence, and of the conflict between modernity, and the conservative traditional value. Whereas the former emphasise the transparency and accountability, the traditional school believes in the "divine rights" of the village administrative hierarchy. For the latter, the misuse of communal funds should be overlooked and should go unquestioned; this trend of affairs definitely set the pace of collapse of this ambitious project. This temporary stagnation marked the anticlimax of a success story.

It became abundantly evident that if the revival of this village water development scheme were to rely on the continuous support of the village community and that of the external elites, this project could not have been revamped because of the fear of the re-edition of the previous financial mismanagement which surfaced very prominently. For these reasons of fear, financial lethargy and the mismanagement of vital and scarce resources, the project stood still. Fortunately, Plan International pumped in the huge financial resources and completed the phase two of the protracted village hydrological project nearly 20 years after the projected date of completion. The lessons from this project are vital if we continue to bank on self-reliant development as a good and rational way forward.

Significance of the Water Project

Perhaps the words of the Ancient Mariner from Coleridge's verse better describe the significance of polluted streams after heavy flooding in this volcanic mountainous district. "Water, water everywhere, not any drop to drink." More than 95% of the population depended on water from streams. But after heavy rainfall, there is a high degree of contamination because much of the alluvial material from the hilly region, the rill and rain-wash from cultivated farm lands eventually find their way into the streams. This contamination accounts for the brownish colour of the streams during the peak rainy season. Under such a state of pollution, drinking water is scarce except for rain water harvested from roofs. It is in these circumstances that one can conveniently borrow the words of the Ancient Mariner cited above.

Moreover, the bulk of the population is concentrated downstream where contamination from human beings or pollution up stream by cattle is high. Against this background, the point emphasized is that neither the few wells nor the polluted streams which had hitherto been used had water of drinkable quality. But there was however enough water for cattle and some domestic purposes.

Society today has constantly changing values which makes it imperative that many rural settlements possess a good water supply. Apart from flooding in early times, man used streams with care. With smaller populations at the time which were widely scattered within the village administrative wards, the small degree of pollution was effectively dispersed by dilution over long distances before receiving further wastes downstream. This situation is expressed by R.H. Wagner (1971) who said that "by the time waste water from a village of a few hundred people had been carried a few miles downstream, the water had been purified naturally – at least, enough to be drinkable in another village." But as the population of the village quarters grew bigger, the greater quantities of pollutants discharged into the stream could no longer be effectively purified by dilution. So with increasing population in recent times, and the lack of concern for the consequences of pollution, man is more than ever before exposed to the risk of water-borne diseases.

Poor and distant water supplies have far reaching repercussions on the population. It does not only affect health but also the total economic productivity. The latter has been used in the sense that the total number of man-hours devoted to fetching water whether by children or women reduces agricultural productivity and other forms of economic output. The contributions of this pipe-borne project from 1984 in terms of time gained which could be profitably diverted to the other sectors of the economy is illustrated by Fig. 4. In the quarters already served, with pipe-borne water, children presently can spend more time on their home work and school exercises while women can increase their farm outputs. The diagram shows that the number of man-hours involved in fetching water varied depending on how distantly located, on the interfluves, the family is from the stream. The nearer the family to the source of water, the fewer the man-hours lost per week in fetching water.

In this rural settlement, man and beast formerly converged for water from the streams. Since there is a much higher population concentrated downstream, the inhabitants drank the dirty water in which people had washed or polluted upstream. Moreover, for several decades, most of the upstream zone has remained a grazier area where the Bororos roam the savannah with their cattle. So contamination from cattle has remained almost an everyday affair.

Prior to the inauguration of the pipe-borne water supply, however, most families went out to fetch drinking water before dawn, when the river was least contaminated as man and beast, who are the agents of pollution, were still at rest. This pipe-borne gravity water supply has not has not only provided the necessary answer to the long standing problem of river pollution on the one hand, but also the distantly located sources of water supply. With the present innovation for this village community, a safe water supply has been ensured for most quarters, and water has been brought nearer the population concentration so that the energy formerly devoted to fetching water for the household could now be effectively directed elsewhere.

Fig. 4: The Time (Man-Hours) Relationship between the Pipe Borne and the Pre-Pipe Borne Water Period in Babungo (Ndop)

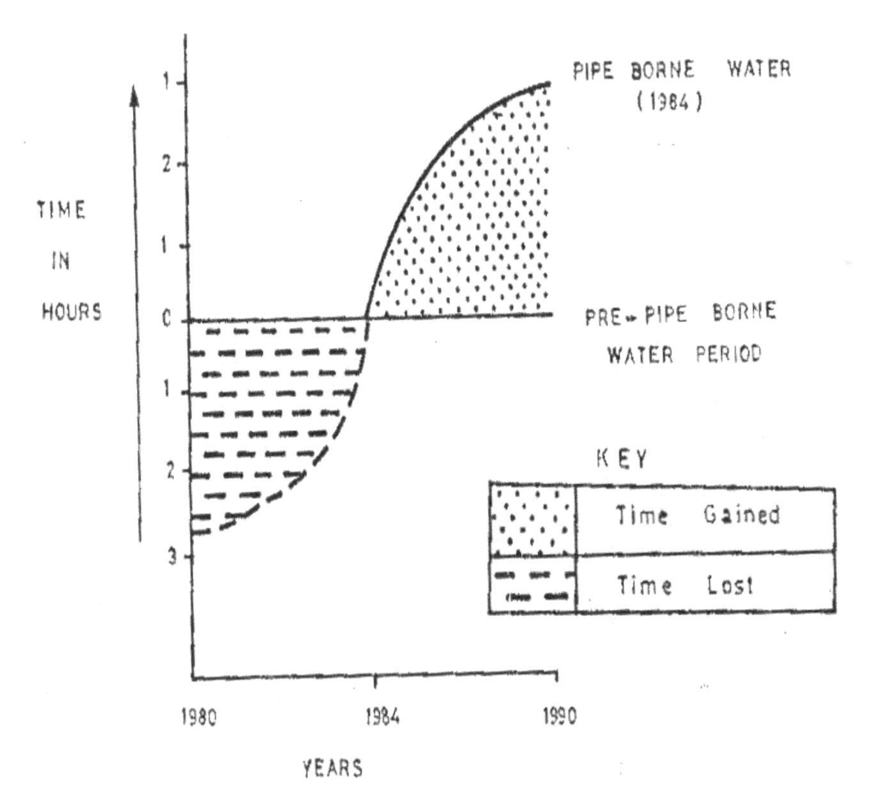

The availability and proximity of an adequate supply of good drinking water can be regarded as one of the quasi effective indices for measuring the standard of living. The World Health Organization (WHO) gives the radius of reasonable access to a water supply point as 200m away. But in Babungo, only a small proportion of the rural population meet the WHO standard since more than 70% of the citizens live within a radius of 0.5-1km from their sources of water supply. The scarcity of this basic human requirement in the home renders the daily curricula pretty tough as the going is rough for those who shuttle between the sources of water and the home.

In most cases, this time-honoured burden of fetching water fell on children and women (Fig.5). With the advent of the village pipe-bore water, the burden has been taken off their necks. The time gained in this direction can not be quantified; its relevance is seen in the relief it brings.

Fig.5: Women and Children Congregating for Water at One of the Village Water Taps

A long time dependence on polluted streams meant that there was a long exposure to water-borne diseases such as schistosomiasis and filariasis. And those who fell victims to the high risk of water related diseases were children According to the 1992 Environment World Report, "The use of polluted water for drinking and bathing is one of the principal causes for infection by diseases that kill millions and make more than a billion people sick each year." So unless the scourge of water-borne diseases such as cholera, typhoid and schistosomiasis are suppressed through the provision of good drinking water, the then slogan "***Health for All by the Year 2000***" had remained a mere illusion for these rural inhabitants even through the turn of the Third Millennium. Health amelioration does not only lie in the mere provision of good water; it can also be seen from the perspective that it provides underground sewage facilities which often require large quantities of water supply. Many modern houses with running water and toilet systems are springing up. Washrooms and bathrooms are becoming fashionable in the countryside. And for a peasant society like Babungo, the cheap and free for all water scheme is a welcome relief as only few might have been able to withstand the high rates of the Cameroon National Water Corporations. (SNEC).

Perhaps, the remarks of J. Bharier (1978) on the importance of easily available and safe water supply from studies in Malawi are analogous and very significant for this rural settlement in the North West Province of Cameroon when he pointed out that:

> Without adequate and convenient supplies of safe water for drinking and washing, people can not attain a reasonable minimum level of existence. Water which is unsafe for human consumption carries and spreads diseases; water which is inconveniently situated leads to considerable loss of productive time, especially for women who have to travel long distances to fetch it, and inadequate supplies of water can frustrate attempted improvements in many aspects of social welfare.

Following the example of the Babungo water project, a few villages in this part of the of the Upper Nun Valley embarked on building their own water supplies through self-help programs or with the aid of some Non-Governmental Organisations (NGOs). Like Babungo, which is located at the foot of the High Lava Plateau, Bamessing for example, also runs a gravity pipe-borne water project. The slopes of the Sabga High Plateau provide excellent source areas for water that requires no treatment. Furthermore, five other village water projects came into operation, namely, the Balikumbat, the Bafanji, Ndop urban area, Bangolan (SCANWATER) and Baba I. Perhaps, the important significance of the Babungo gravity water project was the drive which it generated for similar development schemes in the neighbouring villages in the Ndop Plain of the Upper Nun Valley region of the North West Province.

However, the Scan Water Projects (SWP) of Bangolan and Baba I suffered serious setbacks because the schemes *per se* were high technology projects. Because of their high-tech nature, the maintenance inputs required for the sustenance were not usually available. Hence, the pumps could not function without adequate and proper energy sources.

Future Perspective

This gravity water project has been an ambitious scheme in which phase two which was scheduled to be completed by 1984 was to provide water to all the accessible and heavily populated administrative wards of the village. Ten years afterwards, this objective was only partially attained because the distribution was not as extensive as it appeared in the project description. Since the bulk of the project pivoted on the participation of the entire community, there was considerable reluctance towards communal work from those whose administrative wards or quarters were not yet served with water. Those who were already served by the pipe-borne water felt reluctant contributing to the continuation of the water project. This became a double administrative dilemma, a situation which caused the stagnation of this well-intentioned and beautiful development project. It also became obvious that to maintain a permanent success-story for this village development scheme, executive members of the local branch who are charged with overseeing the progress of the project, would have to avoid all forms of misappropriation of funds since such financial mismanagements in the past, unfortunately 'suffocated' local participation and self-help enthusiasm and initiative.

Furthermore, consumption at the completion of the project was estimated at 130 litres of water per person per day. But during the peak of the harsh 1986/87 dry season, and even into the Third Millennium (2004/2005) most of the water usually drained to low altitudes by gravitation pull so that most of the high population concentration zones received water only in the morning periods. From this experience therefore, it became abundantly obvious that more catchment springs were needed to reinforce those which were currently in use and which had hitherto already proved to be grossly inadequate. This is why Plan International, a Non-governmental Organisation, stepped in to revive the village community water project and used new catchment areas which provided adequate water supplies such that it was able to serve those distally located quarters as Toh Saji, Mbungwi, Tavendong, Ibau and Ibia which were far from the main distribution pipes of the initial village water scheme. Furthermore, the availability of water in the formerly stressed areas added to the success story of this gravity water project.

Perhaps an important lesson to learn from the initial lethargic attitude of the population to pay for the development of the village pipe-borne water project was the concept that water is a natural resource and thus a gift from God, and so it ought to be free. It was only with time that the people understood the necessity to pay for the initial establishment which did not imply the commodification of the good and reliable access to potable water. This project has, no doubt, helped to empower our rural communities to identify, plan and execute village water schemes (Mills & Turusbekov, 2008) and other development projects in collaboration with the government of Cameroon and Non-governmental Organisations. Such community participation has thus enhanced village ownership of the hydro-schemes thereby blocking the entry of the Cameroon National Water Corporation (SNEC), which in some cases, could step in to put metres and bill the inhabitants under the pretext that it is purifying the water. This community project improved the access to safe drinking water for thousands of the rural population.

References

Bharier, J. (1978): Improving Rural Water Supply in Malawi, *Finance and Development, Quarterly Publication of IMF & the World Bank*, September 1978, Vol. 15, No. 3, pp 34-36

Drinking Water – A Basic Necessity of Life. *The Courier ACP/EEC, N° 96*, March-April 1986, pp 62-96

Funke-Barrtz, M. (2008): Water for all–but how? *Development and Cooperation (D+C)*, Vol. 3, 2008:1, pp 34-35.

Gondwe, E. (1986): Planning Water Supply in Tanzania, *The Courier ACP-EEC*, No. 96, March-April, pp 95-96

Mills, M. & Turusbekov, Esen (2008): Making Water Flow Uphill, *Developments Issue* 41, 2008, pp 30-31.

North West Development Authority. (Mission de Dévéloppément de la Province du Nord Ouest (MEDINO).

Progress Report of 30th November on the Babungo Water Supply Project, Community Development department, Ministry of Agriculture, Republic of Cameroon

Swiss Association for Technical Assistance. A Water Company for Rural Development attached to the Department of Community Development

"The Rime of the Ancient Mariner," The Lyrical Ballads by S.T. Coleridge

Wagner, R.H. (1971): Environment and Man, W.W. Norton & Company, Inc., New York, pp 107

World Development Report. The World Bank, Washington D.C., 1980, pp 75

Chapter Ten

The Dilemma of Rice Stockpiles: The Case of the Upper Nun Valley Development Authority of the North West Province of Cameroon

Summary

The production of rice in the Upper Nun Valley Development Authority which was undertaken to meet part of the domestic demand growth rate of 3.5% and also to ameliorate the standard of living of the rural cultivators, was faced with the problems of stockpiles resulting from unmarketed quantities of rice. The ineffectiveness of the quota system, the fraudulent importation of large quantities of duty free rice into the country, and the non implementation of government policy accounted for these stock piles. The state in the circumstances did not seem to have adequately protected the indigenous corporations by closing the doors of the Cameroon market to external competition. Rice cultivation in Ndop Plain flourished from its inception; but around the 1980s however, there was a slow down in production because of unmarketed stockpiles of this cereal.

Attendant consequences from the rice stockpiles at that time probably comprised of rural exodus by the youths, a fall in the rural income generating capacity, and a much slower modernisation of the country side.

Introduction

Prior to the early 1980s, the UNVDA was lauded for increased indigenous rice production in Cameroon. But this production did not bring with it the usual financial euphoria that innovations and success stories dictated because of the huge unmarketed rice stockpiles in warehouses, particularly in the Upper Nun Valley of Cameroon. At a time when most African countries were stricken by the pangs of hunger, Cameroon enjoyed a remarkable level of food sufficiency and remained Central Africa's granary. And what was more: the chronic food insuffiency which has plagued many countries of the continent contrasts with food stock piles in some parts of Cameroon. Rice provides a case in point for the SEMRY[1] project of Yagoua in Nortrh Camrroon, the SODERIM[2] of the Mbo Plain in the West Province and the UNVDA[3] of Ndop, Ngoketunjia Division. The latter, UNVDA, is one of the parastatals charged with agricultural developments which concentrated on the introduction and cultivation of rice in the fertile Upper Nun Valley in the North West Province with its headquarters at Ndop (Fig.1). The target of this corporation was to develop 3000 hectares of land by 1985 using a total labour force of 6000 farmers or households and this objective was achieved as scheduled since most farmers embraced this new agricultural innovation with plenty of optimism for a better tomorrow.

Fig. 1: Location of Ndop, Ngoketunjia Division Seat of the Upper Nun Valley Development Authority

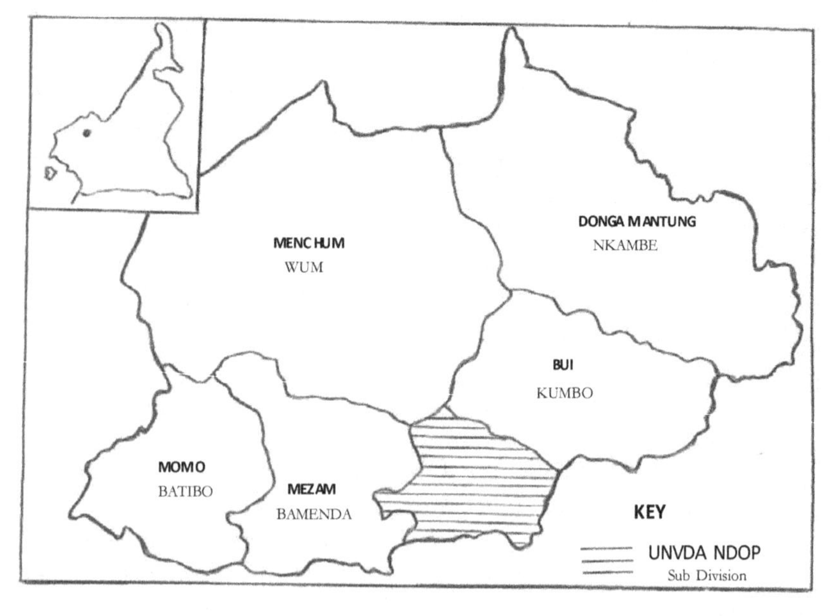

Realising that the national consumption of rice was estimated to rise from 52,000 tons in 1980 to 90,000 tons in 1985, and that this consumption was projected to be about 200,000 tons by the year 2000, government saw the pressing need under the Fifth Year Development Plan for improving rice production, processing, packaging and distribution. Thus confronted with a domestic demand growth rate of about 3.5% during the 1981 to 1986 period for the national territory, the three corporations were encouraged to boost production to meet up with the increasing demand. This increase in production did not go without its own problems.

Methods and DATA SOURCES

Following the persistent discussions of the poor financial status of the UNVDA and the fall in its economic activities, we examined from a historical perspective, the causes of the slow down of this corporation. Using production and marketing data of this corporation and from interviews of the farmers and the UNVDA administration, it became obvious that the rice stockpiles were the fundamental cause of the financial instability of this corporation that shook it to its very foundation or roots.

We examined the causes of these stockpiles and their economic consequences especially on these small holders who produced the food commodity. Furthermore, this study was undertaken with a view to suggesting possible remedial measures for the future if the Upper Nun Valley Development Authority had to survive as a viable corporation at the service of the peasant rural communities. In this direction therefore, a number of relevant questions were worth examining. The first of these was whether the then stockpiles were due to increased production or a bumper harvest. Secondly, deriving from such stockpiles, another cardinal question was whether the indigenous agro-industry was adequately protected by government against undue foreign competition by South East Asian countries which produced rice much more cheaply because of intensive mechanisation. The final question addressed the effectiveness of the quota system (systeme de jumelage) which stipulated that rice merchants had to buy fixed percentages from the indigenous corporations and the rest imported from abroad. These and other related issues probably provided useful pointers for the stockpiles in question in the Upper Nun Valley Development Authority.

Causes and Solutions for Stockpiles

Price discrimination, the ineffective quota system, indirect "smuggling" and the non-implementation of government policy by Cameroonian rice merchants-all did combine to influence the stock piles of rice of the Upper Nun Valley Development Authority in one way or the other. Production certainly was rising but remained within the projected figures and should thus have constituted no problem to worry about, everything being equal.

Given the constantly increasing production trend of rice in Cameroon (Table 1 & Fig. 2), government should have issued yearly importation licences for a correspondingly smaller percentage of foreign rice so that the home produce should have a wider market. The estimated paddy production for the 1985/86, for example, was put at 129,000 tons, corresponding to an annual increase of 18%. For Cameroon to cope with the increasing production of rice from the three corporations in the country, it was most urgent that the quota system for home produced and foreign rice be revised almost yearly based on projected production figures. This idea of trends in rising production was also corroborated by figures of previous years of production as illustrated by Table 2 below.

A brief survey of the production trend indicates that "results obtained in the field of rice culture were encouraging"[4] for the traditional sector (small holders) and the corporations. Such increasing yields should have effectively signalled the drastic reduction of imported rice or the partial closure of our doors to foreign rice. This austerity measure should have been taken if the farmers had to maintain faith in the state in spite of the credit, seeds and other agricultural inputs which the government and corporation at the time made available to the farmers.

However, if the quota system which was evidently down-trodden by some dishonest merchants who did not realize the harm they were wrecking on Cameroon's economy, were scrupulously examined and rigorously implemented, such a deplorable marketing situation would have been forestalled in the years ahead. According to the terms of the quota system (system de jumelage) established by the state, all rice importers were obliged to buy a certain quantity of local rice from the indigenous corporations. To get around this clause and still obtain the tonnage required, the merchants imported more rice under the pretext of re-export to the neighbouring countries of the Communauté Economique et Monétaire d'Afrique Centrale (CEMAC)[5] or the then countries the Union de Développement Economique d'Afrique Centrale (UDEAC) and obviously, these quantities do not attract any custom duty.

Table 1: Production by the Three Rice Producing Projects in Cameroon

Corporation	1981/82	1982/83
Semry	53,639	74,026
Unvda	3,160	3,500
Soderim	600	800
Total in Tons	**57,399**	**78,326**

(Adopted from Food self sufficiency in Cameroon/Bamenda Agro-Pastoral Show 1984)

Table 2: Basic Trends in Paddy Rice Production in Tons from 1981 – 1986

1981/82	1982/83	1983/84	1984/85	1985/86
49.320	72.525	97.150	91.000	129.000

Source: Ministry of Agriculture

The duty-free status on such high tonnages of imported rice permitted the merchants to sell the imported quantities far below the stipulated government prices and still bagged substantial gains. Here, it seemed abundantly obvious that there was no follow up of state policy governing the importation of foreign rice.

So another serious problem that confronted this agro-industry was "smuggling" because some merchants obtained importation licenses for re-export of this cereal to our neighbouring land locked UDEAC[6] countries like Chad and Central African Republic. But much of this rice landed somewhere else in the country so that it never got to its officially purported destinations. What this actually meant was that Cameroon imported more than the real quota that was supposed to supplement the home production as estimated by government. This anomalous situation was also corroborated by the economic and Financial Report for 1986/87 which stated that:

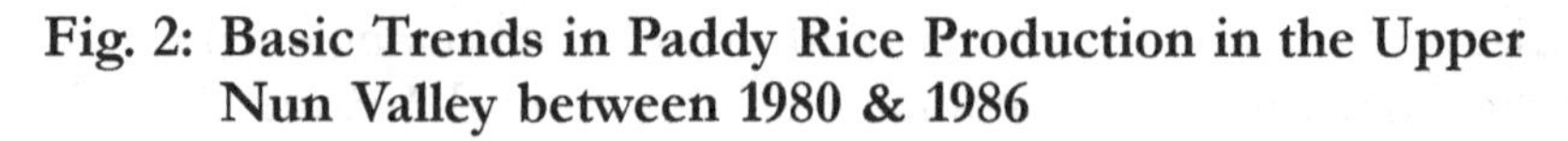

Fig. 2: Basic Trends in Paddy Rice Production in the Upper Nun Valley between 1980 & 1986

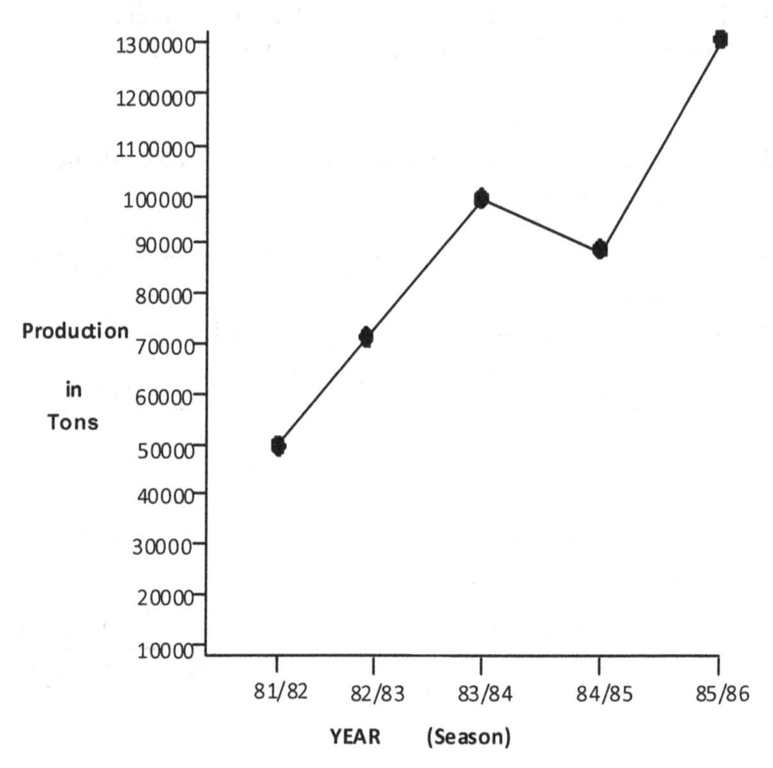

> "The chronic slump in local rice owing mainly to its very high cost price coupled with very often fraudulent imports of this product (fraudulent clearance at the port) instead gave rise to an abundance of rice on the market."[7]

Since the imported rice sold at a lower price, the consequences were that the indigenous corporations began to have stocks of unsold rice. For 1983/84 alone, Cameroon was stuck with large quantities of unsold rice since as much as 126,000tons of imported rice had entered the country whereas national consumption was estimated at 90,000 tons.[8]

In the face of such clandestinely increased tonnage of imported rice into the country, the merchants sold at very marginal prices and still made enormous profits since their relatively lower prices attracted a much wider market than the home-produced rice from UNVDA, SEMRY and SODERIM. As alluded to earlier, the lower prices for this cereal on the whole were facilitated by the duty-free fraudulently landed quantities. So the duty-free aspect remained one of the crucial factors around which the stockpiles of the indigenous corporations revolved.

From these floods of cheap rice, the internal market drastically shifted from home cultivated rice as a result of price discrimination. At that time, the Ministry of Commerce and Industry set the price of one kilogram of paddy at 78 FCFA, and by the time marketable rice left the factory, it was 160 FCFA per kilogram (Table 3).

Table 3: The Situation of Stocks by the 25th October 1986 Valued in Mercurial

Products	Quantity (In Kg)	Unit Prices (In Fcfa)	Value (In Fcfa)
White Rice (Threshed Rice)	325,041	160	52,006,560
Broken Rice	1,100	60	66,000
Flour	20	40	800
Bran of Haystack	0	25	0
Bran of Shells	1,950	25	48,750
Total Cost of Finished Products			52,122,110
Paddy For Factory Processing	6,012,958	78	469,010,724
Total Value In Stock in Mercurial Terms			**521,132,834**

(Adopted from UNVDA Technical Section)

Since cheaply produced imported rice was barely 100Fr per kilogram as compared with 160Frs for locally produced rice, the drift towards foreign rice stocks was almost pretty effective over the national territory. These problems of higher prices had been compounded by additional transport cost for conveying the rice to consumption centres like Bamenda, Douala and Yaoundé.

The comparative costs of imported and locally produced rice are given in Table 4 below. A Yaounde based survey showed that rice was marketed at prices which were far below those stipulated by government. The evolution of prices of imported rice for the 1982/83, 1983/84 and the 1984/85 seasons were 12,577Frs, 13,135Frs and 14,909Frs respectively. But the actual market price for 92 kg of imported rice up to December 1986 was 11,000FCFA instead of 14,909FCFA for the 1984/85 season (Table 4). These prevailing market prices were not only 35.5% less than the government value, but even far below the 1982/83 level of 12.577FCFA. Such an alarming drop in the prices of imported rice was not fortuitous, and could be justifiably explained in terms of the huge rackets derived from the sale of the duty-free commodity that flooded the Cameroonian market.

Following the corporation's huge stockpiles of rice, the authorities of the UNVDA devised a new commercialisation strategy aimed at capturing part of the market. In this regard, the prices of locally produced rice were comparatively reduced to match those of imported rice.

Table 4: Comparative Costs of Imported and Locally Produced Rice

Quantity	Imported Rice	Local Rice	Produced New Cost
90kg	-	12,500fr	9,000fr
50kg	6,500fr	6,750fr	4,500fr
92kg	11,000fr	-	-

(Market Survey, 1986)

These proposed new prices are indicated in Table 3 above. Since a lot of customers drifted away because of the comparatively higher price for the home-produced rice, it was hoped that this reduction was going to attract more consumers. Even though the proposed price reduction appeared to be one of the long term concrete solutions to the problems of stockpiles, the lack of publicity to this effect meant that the locally produced rice did not still get down to the ordinary consumer who discriminated against it purely from the differential price perspective.

Furthermore, in a bid to help the corporations clear their stocks of rice, the Ministry of Trade and Industry imposed a two month ban on the importation of rice, a ban that was lifted on the 8th of

October 1986. But what difference did this make when the Cameroonian markets were already flooded with tons of imported rice. In fact, the two month period was probably just enough time to enable the merchants to clear off the excess stocks of the imported rice which they had. What this implied in concrete terms was that the rice agro-industry was not sufficiently protected against foreign competition. Furthermore, the following statement clearly shows that the state was fully aware of the problems of the corporation, and the stockpiles suggest that the problems were never combated with the vigour they deserved.

> "At the end of the 1985/86 financial year, production was expected to be about 105,200 tons (paddy rice) and this was going to aggravate the storage and sales problems faced by the companies involved in the development of rice cultivation."[9]

It would therefore seem that stricter measures should have been taken in order to ensure the implementation of the laws governing the importation of rice since these laws had up till then been rather cosmetic.

Consequences Of Stockpiles

Cameroon with all its organs of rural development took a long time to encourage and sensitise the farmers that rice cultivation could provide an alternative or additional source of income for the rural population. These measures included the payment of attractive prices to farmers for their paddy by the corporation, the provisions of loans to rice farmers, the construction of irrigation systems or drains, the provision of new strains or improved seed varieties, and above all, the constant availability of technical advice and supervision. Another significant contribution of the agro-sector to rural development was the construction of access roads and farm-to-market roads in otherwise economically neglected areas. These efforts were, indeed, enormous. And when this agricultural innovation for this one-time alien crop culture was accepted, it was then very necessary that the spirit be rekindled or kept alive. If these small holders of the Upper Nun Valley lost the impetus, the results, it seemed, were likely to be most negative for the future of rice production in this part of the country.

The first was that the lethargic attitude of disappointed farmers could work like a canker worm in discouraging prospective young farmers as the universal argument would be "why produce rice when there is no market, and why invest money, labour and time when there would be no financial returns." This was cheap but vicious propaganda which could have been used to derail some marginal farmers.

Moreover, this governmental crusade towards the improvements of the socio-economic conditions of the rural poor should be looked upon not only in terms of the financial remuneration derived directly by the farmers, but also the fact in guaranteeing food sufficiency, Cameroonians ate much better in the 1980s in a continent where many countries did not even succeed in keeping hunger at bay. This food sufficiency could also be looked upon as a measure of Cameroon's effort in the domain of rural development and production, and the contribution of the country-side in the economic growth of the nation.

Were Cameroon and Africa to eliminate the problem of food redistribution, areas of such stockpiles could serve as valuable source areas of food for those countries of the continent suffering from chronic food deficits that arise from either the occurrence of draught or other natural disasters. With effective redistribution, Cameroon which attained self-sufficiency in food could play a vital role as an important bread basket, by "advancing beyond national self-sufficiency in food, since we have the possibility of making our country one of the main granaries of Africa."[10]

As rice for the previous season remained unmarketed, a number of consequences became apparent. The most important was that the corporation was not able to purchase new paddy for the 1986/87 year which ran from December to January. Its inability to buy new paddy was primarily motivated by the fact that the corporation was in the red as unsold stock meant that money was tied down indefinitely. At that time, only God alone knew when the marketing situation was expected to improve for farmers to get their overdue payments from the corporation.

By 1986, out of 415,000,000 FCFA which the corporation borrowed from the bank to finance the purchase of paddy and other operations, only part was reimbursed. In the face of mounting debts, it became incredible to conceive the negotiation of further bank loans for the purchase of additional stocks of paddy when there was more than six thousand tons of paddy (6,012,958 kg), an equivalent of 3,600 tons of clean rice still outstanding. In mercurial terms, this was equivalent to 469,010,724 FCFA. Finished produce at the time amounted to some fifty-two million (52,122,110 FCFA). Cumulatively, therefore, the total value of stocked rice was more than five hundred and twenty-one million francs (521,132,834 F CFA). The quantities and values showing the situation of the stockpiles by the 25th of October 1986 are given in Table 3.

Secondly, there was absolutely no room for the storage of further stocks of paddy. Consequently, there was absolutely no sense in buying more paddy when the unprocessed old paddy was wasting as a result of long exposure to the humid climatic conditions of this part of the Upper Nun Valley. If the corporation bought no more paddy from the small holders, there was no problem as they could easily process and market the small quantities they produced. But the problem really arose because farmers grew substantial quantities of rice. Just as the corporation itself could not market its rice, so also was the fate of the more prosperous farmers. The net result of the non purchase of further paddy was the eventual fall in production since no one wanted to produce when there was no available market for the commodity. The fall in production stemmed from the drastic fall in cultivable acreages since there were no proceeds from the previous harvest to finance the cultivation of large farm plots. For those, however, who embraced the cultivation of rice and would continue in spite of the financial problems following the non purchase of new paddy by the corporation, they were more likely to cultivate smaller acreages. Once production fell and farmers got into alternative occupations, it became difficult, if not impossible to revamp their interest in farming since the gloomy memories of the past would probably linger on for a long time. This was a pessimistic view which did not preclude the fact that better future marketing policies could still bring back even more people into the cultivation. But these were mere speculations at that time; however, the prevalence of rice scarcity on our markets today has rekindled that agricultural impetus which we thought better marketing strategies would bring. The inflation trends in our present economy have provided such high prices that rice costs almost three times their 1986 value.

Rice cultivation in the Upper Nun Valley started way back in 1966. Since then, the small holders have cultivated it for 40 long years that they have come into terms with this form of economic activity, and probably developed some sort of agro-inertia for it. In concentrating on rice, many farmers have either abandoned the growth of other crops or relegated them to a secondary position. A collapse in the corporation therefore was not only going to affect the 6000 farmers but 6000 families since these farmers and their families supply part of the intensive labour for rice cultivation. It cannot be overemphasized that the financial returns from the sale of rice played an important role in the well-being of the families. In concrete terms, part of the money went to pay for the education of children, feeding, the modernization of homes and the provision of health and other facilities to the small farmers which were uncommon to the ordinary peasant. These facilities of better housing and better access to basic health care illustrate an improvement in the standard of living of the small farmers – thanks to the agro-industrial innovations of the UNVDA. Moreover, apart from the 6000 farmers, about 700 direct staff of the UNVDA such as demonstrators, office and field staff and temporal workers were affected by retrenchment.

Unlike the farmers of SODERIM of the Mbo Plain who diversified their incomes through the raising of pigs and poultry, most of the 6000 farmers of the Upper Nun Valley practised only rice monoculture for a long time; so they had nothing substantial to turn to when the home-produced rice faced difficult internal marketing problems.

A brief survey of labour force showed that few elders were engaged in this agro-business. Most of the cultivation was in the hands of young people who were attracted by the ever increasing financial returns from rice. Once they packed off from the soil because of stock piles the tendency was the migration into the urban areas which offered better prospects for employment. This attraction of the youth by the urban economy remains a strong force to reckon with in our townships today. Rural exodus of this nature does not only slow down the pace of development of our country side, but also threatens our present security in food self sufficiency which implies freedom from hunger, malnutrition and misery.

Conclusion

In the search for possible remedial measures to combat stockpiling as well as ensuring the growth and survival of UNVDA and the other rice corporations, it seemed most appropriate that there should have been a vigorous pursuit of government policy governing the importation of rice. And in order to forestall a stagnation of our rice produce as unmarketed commodity in silos and warehouses, the doors of the Cameroonian market should have been closed to foreign rice imports for a sufficiently long time to enable the corporations clear off their existing stocks. Furthermore, the then price reduction proposed by the corporation as a means of combating stockpiles was an important move in the right direction since the consumers discriminated against indigenous rice because it was more expensive.

As rice on the Cameroonian market as at 2008 has become a highly demanded and expensive commodity because of the recent price hikes, the current market situation should therefore serve as an incentive for the intensification of rice cultivation in the Upper Nun Valley of Cameroon. In the circumstances, production trends clearly indicate that the UNVDA and its host of associated farmers have moved from a period of economic depression characterised by rice stockpiles to an era of economic boom or renaissance of rice shortages characterised by price hikes.

Notes

1. SEMRY: Corporation for the Expansion and Modernisation of Rice Cultivation.
2. SODERIM: Mbo Plain Rice Development Corporation.
3. UNVDA: Upper Nun Valley Development Authority.
4. Food sufficiency in Cameroon: Bamenda Agro-Pastoral Show.
5. Communauté Economique et Monétaire d'Afrique Centrale (CEMAC)
6. Economic and Customs Union of Central African Countries (UDEAC).
7. Economic and Financial Report, Ministry of Finance, Finance Bill for 1986/87, TR. 86/419, pp 65.
8. Cameroon Tribute, of 13th December 1984; cameroon's Food self-sufficiency, Results of good planning, pp 19.

9. Economic and Financial Report, Ministry of Finance, Finance Bill for 1986/87, TR. 86/419, pp 27.

10. Food, Forestry & Environment: The Challenge to Rural Poverty in Africa, Fifth World Food Day & 40th Anniversary of FAO, Douala, Cameroon. June 1986, pp 16.

Chapter Eleven

Cropping Intensity and Post-cultivation Vegetation Successions: Developing Sustainable Agro-ecosystems in Ndop Plain, Cameroon

Summary

The sustainability of traditional farming systems in many parts of Africa is threatened by losses in the variety of species, reduction in land, forest, soil and water resources under demographic pressure. Together with foreign influences these farming system are no longer in equilibrium with local culture and ecology and are therefore disintegrating due to the lack of local capacity to adjust to these changes. This has led to environmental degradation. The paper employs a combination of qualitative and quantitative techniques to: analyse the farming systems, assess the effect of cropping intensification on the climatic climax vegetation, and to establish the post-cultivation vegetation successions in fallows. It reports four crop fallow rotations: crop 3 years with 1 year bush fallow, crop 5 years with 1-2 year bush fallow, crop 10 years with 1-3 year bush fallow and continuous cropping. The post-cultivation successions present a plagioclimax dominated by *Hyparhenia* species with scattered shrubs maintained by cycles of cultivation and burning. The paper concludes that these shifting cultivation cycles involving cropping intensification and post cultivation savannization have deprived the farming system of its ecological benefits, that is, nutrient recycling through a long and mature fallow. The inability of the vegetation to reconstitute itself requires the development of farming systems that combine trees and crops in the field because of proven micro-ecological and ecological benefits. The paper therefore identifies the scope for the development of socially, economically and ecologically sustainable agro-ecosystems as a development path for shifting cultivation systems.

Introduction

The sustainability of agriculture in many parts of Sub-Saharan Africa is threatened by losses in the variety of species, reduction in land, forest and water resources, soil erosion, salinization, acidification, desertification and environmental pollution (Bassey and Ndenecho, 2006). Traditional farming systems which over centuries developed in constant interaction with local culture and local ecology have disintegrated because of the lack of local capacity to adjust to population growth and the influence of foreign values. The inability to mange change has led to severe environmental degradation. (Lawton and Wike, 1979; Weiskel, 1989; TAC/CGIAR, 1988). In response to these influences there has been a tendency towards intensification characterized by short duration fallows with low external inputs (Chambers *et al.*, 1989; OTA, 1988). For most of Africa, production therefore lags behind population growth. As new technologies to intensify land use in a sustainable way have not been developed or are not known to farmers, they are often forced to exploit their land beyond its carrying capacity (Harris, 1999; Sachs, 1987). The over use of land under demographic pressure and the expansion of farm boundaries and small holdings to new, often marginal farming areas leads to deforestation, soil degradation and increased vulnerability to torrential rains and droughts (Reijintjes *et al*, 1992). Many African land use systems are therefore in the midst of such downward spiral of nutrient depletion, loss of vegetation cover, soil erosion and economic, social and cultural disintegration.

The paper seeks to analyse the farming systems in both qualitative and quantitative terms, and to assess the role of cropping intensification on the climatic climax vegetation, and post-cultivation derivatives as they relate to the development of sustainable agro-systems in peasant communities.

The Study Area And Study Sites

Ndop plain is an intermontane basin in the Bamenda Hyghlands (Fig. 1). The average attitude is 1200 m above sea level. It has a humid tropical climate with annual rainfall in the range of 1500 to 2000mm. The wet season lasts from mid-March to mid-November. The rest of the monthsare dry. Annual average temperature is 21.3^{O}C. Large water deficits are experienced from December to February.

Fig. 1: Location of the study area (Ndop) and villages investigated

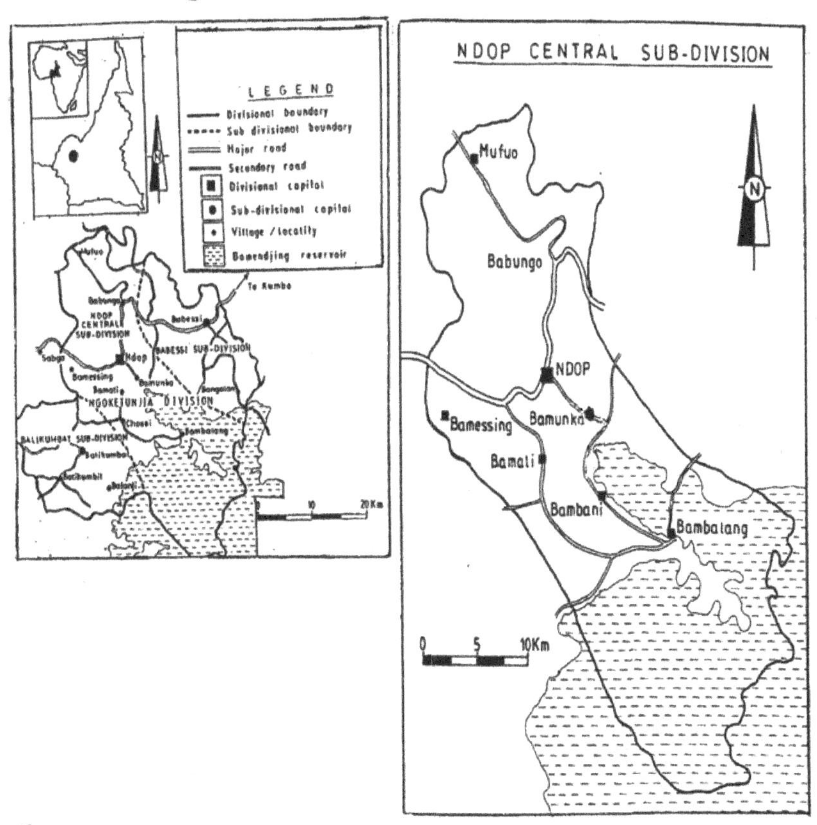

Three geological formations characterize the area: granites and migmatites in the north, ancient basalts in the south, and alluvial deposits in the centre and east. The main pedomorphic units are: alluvial soils on the Nun River flood plain, alluvio - colluvial deposits on the gently undulating foot-hills in the north, soils developed on basalts and trachytes in the south and fine-grained granitic soils on undulating hillocks in the plain (Kips et al., 1987) The vegetation is mainly wooded grassland with gallery forests at the head of thalwegs and *Raphia Vinifera* palm bushes in swampy areas. The natural rain forest vegetation has greatly been disturbed by man (Hawkins and Brunt, 1965; SEDA, 1983; Champaud, 1973). The average population density is 96 inhabitants/ km^2.

Demographic pressure on land is high resulting in an average farm size of 1.54 ha fragmented into two or three plots at different locations in the village. Shifting cultivation with short duration fallows is typical. This has resulted in the need to procure external inputs and to modernize the farming systems (Lambi, 2001; Mchugh 1988; Aseh, 1997). The average farm family size is 6 with 3 active farm workers. The main field sites investigated were (Fig. 1):

- √ Nun flood plain south of Bamuka village chief's palace: cultivation fallows and fields on silty-loams.
- √ Colluvial zone west of Rest House along the Bamenda-Kumbo Highway; cultivation fallows and fields on granitic colluvium.
- √ Colluvial unit in Babungo Agricultural station: cultivation fallows and fields on lava colluvium.
- √ Fine – textured granitic soils in Bambalang: forest remnant and regenerating fallows and cultivated fields.
- √ Coarse-grained granitic soils beyond Bamali village chief's palace: fallow plots and cultivated fields.

Methods and Data Sources

In order to establish the cropping system, cropping intensity and cropping sequence pattern 130 randomly selected fields were investigated in the 5 villages. The frequency of selected crops in crop associations and the distribution of crop associations on 23 randomly selected fields were the main aspects of the cropping pattern investigated with the assistance of the various village agricultural extension workers. The results were complemented by the maize – based farming system described by McHugh (1988). Each farmer was interviewed on the cropping sequence, cropping years and duration of fallows. Once the duration of fallows was established 5 post-cultivation vegetation successions per fallow year were investigated using random quadrant sampling. The distribution or the way plant species were dispersed over a quadrant was done by the inspection and identification of species present or absent from the area within each quadrant. The species so identified were then grouped under three abundance classes per fallow year:

Class 1: very numerous, Class 2: numerous , Class 3: not numerous. The data obtained from the farming system analysis and quadrant sampling were used to develop a model of the shifting cultivation cycle, the process of cropping intensification and post – cultivation vegetation successions. The model assisted in the identification of the scope for the development of a sustainable agro-system under shifting cultivation systems with short duration fallows.

Results and Discussions

The cropping systems can best be described as multiple cropping systems with mixed inter-cropping, that is, two or more crops are grown simultaneously with no distinct row arrangement and row inter-cropping, that is, growing two or more crops simultaneously with one or more planted in rows (Table 1).

Table 1: Observed frequency of selected crops in crop associations in Ndop plain

Crop	No. of fields	% of fields in which observed
Maize	130	100
Colocasia	86	66.1
Macabo (*Xanthosoma*)	83	63.8
Groundnuts	82	63.0
Okra	73	56.1
Yams	72	55.3
Beans	69	53.0
Pumpkin	61	46.9
Plantain	57	43.8
Egusi melon	56	43.0
Cassava	42	32.3
Bananas	37	28.4
Sweet potatoes	20	15.3
Cowpeas	17	13.0
Huckleberry	16	12.3
Coffee	16	12.3
Raffia palm	7	5.3
Oil palm	7	5.3
Mango	7	5.3
Bambara Groundnut	4	3.0
Irish potatoes	3	2.3

Mean number of crops per field = 7.4 Standard deviation = 2.0 Range = 2-14

Table 2: Distribution of crop association on fields in Ndop Plain

Crop association	No. of fields	(%)
Four crop associations:		**(12%)**
Maize + Groundnuts + Cocoyam** + Egusi	1	(4)
Maize + Groundnuts + Beans + Egusi	1	(4)
Maize + Groundnuts + Okra + Egusi	1	(4)
Three crop associations:		**(12%)**
Maize + Groundnuts + Cocoyam	4	(17)
Maize + Groundnuts + Bean	1	(4)
Maize + Groundnuts + Cowpea	1	(4)
Maize + Groundnuts + Okra	1	(4)
Maize + Groundnuts + Cocoyam	1	(4)
Maize + Groundnuts + Egusi	1	(4)
Two Crop associations		**(35%)**
Maize + Groundnuts	3	(13)
Maize + Cocoyam	2	(9)
Maize + Bean	1	(9)
Maize + Yam	1	(4)
Sole crops		**(13%)**
Maize*	3	(13)

* Colocasia (Taro) and Xanthosoma (Macabo)

** One field was pure maize. Two had other crops at low density

Table 2 presents the main types of crop associations. Maize is represented in all associations as the main crop. The various crops presented in table 1 are associated in varying combinations of few crops. The main crop associations observed on farms include:

- Groundnut (*Arachis hypogaea*) is the most common intercrop with maize. It is present in 66% of the maize fields. Groundnuts are planted with and harvested just after the maize. They yield very little because of low temperatures, high humidity, reduced sunshine hours during the wet season, shading by the maize and pests (Rosette and animals).
- Beans *(pliaseolus vulgaris)* are the most common intercrop with maize.
- Cassava (*Manihot esculeta*) is more common in some fields than others, especially in the areas with poorer soils. Cassava was present in 32% of maize fields. It is planted with the maize

but, stays in the field between 1 and 3 years, with maize being planted around it each year.

The cropping pattern sequence varies considerably from farmer to farmer. Some farmers crop the land continuously, growing the same association year after year. Others plant the same association for several years, followed by one or two years of bush fallow. Still others vary the association throughout the rotation, putting crops more demanding on fertile soils. The length of the rotation fallow also differs from farm to farm apparently partly in response to soil fertility conditions and the degree of population pressure and / or the terms of land tenure. A few examples of cop fallow rotations reported are:

- Crop 3 years / Bush fallow (1 year)
- Crop 5 years / Bush fallow (1-2 years)
- Crop 10 years / Bush fallow (1-3 years).
- Continuous cropping.

This certainly indicates the influence of population on fallow durations (which are too short). Second season cultivation of maize is not practised because:

- The incidence of stem borers increases considerably towards the end of the wet season;
- The first season crop is still in the field as late as mid-August,
- The potential for cattle damage is much during the dry season.
- Soils are droughty during the dry season.

The farmer is therefore faced with the problem of increasing crop hectarage both in time and in space. It should however be noted that a second maize crop will further lay pressure on soil resources thereby accelerating soil fertility depletion. Any considerations of a dry season crop must foresee a good input delivery mechanism for the maintenance of water delivery systems, structures and soil fertility.

Hawkins and Brunt (1965) described the climax vegetation of Ndop as lowland rain forest. Based on life form they identified 46 tree species, 4 shrubs, 5 creepers and 4 grass species. The main derivatives observed in Bambalang and Bamali at 1, 260m above sea level were *Syzygium guineensis* tree savanna. *Terminalia* shrub savanna with *Hyparrhenia* grass were also observed on fine-grained granitic soils in the Bambungo area. Today, the landscape is a mosaic of tree and shrub savanna, *Terminalia glauceseens, Lophira lanceolata*, and *Annona senegalensis.* It is characterized by a dense network of gallery forest (Ndenecho, 2005; Ndenecho, 2006). On the fine-granitic soils is a pure carpet of *Hyparrhenia cymbaria, Hyparrhenia dissoluta* and *Hyparrhenia rufa* grassland with tufts of *Clappertonia ficifolia.* The savannization process is the result of pyrogenic and anthropogenic factors. This study focused on post-cultivation successions.

Table 3: Post-cultivation invasion of fallows by grasses and herbs in Ndop plain

Abundance classes	Class 1: Very numerous	Class 2: Numerous	Class 3: Not numerous
1st year fallows	*Erigeron floribundus* *Ageratium conyzoides* *Anisopapus africanus* *Guizotial scabra* *Laggera alata* *Laggera pterodonta*	*Rynchelytrum repens*	*Pteridium aquilinium* (bracken fern)
2nd year fallows	*Imperata cylindrica*	*Erigeron floribundus* *Ageratium conyzoides* *Anisopapus africanus* *Guizotial scabra* *Laggera alata* *Laggera pterodonta*	*Rynchelytrum repens*
3rd year fallows	*Hyparrhenia spp.* *Digitaria spp* *Melinis minuntiflora*	*Imperata cylindrica*	*Erigeron floribundus* *Ageratium conyzoides* *Anisopapus africanus* *Guizotial scabra* *Laggera spp.*
4th year fallows	*Hyparrhenia rufa* *Hyparrhenia dissoluta* *Hyparrhenia cymbaria*	*Imperata cylindrica*	*Imperata cylindrica*
5th year fallows	*Hyparrhenia rufa* *Hyparrhenia dissoluta* *Hyparrhenia cymbaria*	*Digitaria spp.* *Melinis minuntiflora*	*Imperata cylindrica*

Table 3 presents the post-cultivation successions observed in fallows. Slash-and-burn shifting cultivation with short fallow durations of 1 to 5 years provokes a savannization process. The succession of fallow grasses and weeds is as follows (Plagiosere):

- During the first year there is rank weed growth dominated by members of the compositae, that is, *Erigeron floribundus* dominates together with *Ageratium conyzoides, Anisopapus africanus, Guizotial scabra, Laggera alata* and *Laggera pterodonta.* The annual grass *Rynchelytrum repens* is often also abundant on first year fallows while *Pteridium aquilinium* (bracken fern) is rare.
- During the second year of fallow, there is invasion by *Imperata cylindrica* which may become dominant towards the end of the rainy season.
- Third year invasion is by grasses such as *Hyparrhenia* species, *Digitaria* species and *Melinis minuntiflora.*
- Fourth and fifth year fallows are rare. However, in areas where population pressure on land is low, fourth and fifth your fallows show a gradual decline in *Imperata cylindrica* and *Hyparrhenia spp.* dominance. *Imperata cylindrica* tends to persist on land which is under permanent cultivation or is being fallowed from a year to two because it is capable of growing on soils of poor fertility. It is a light demanding species and is shaded out later in the successions by the taller *Hyparrhenia* grasses.

Hawkins and Brunt (1965) report that germination rates of *Hyparrhenia spp.* are very low. This may partly account for the time it takes before it re-establishes itself in old farmlands. When left uncultivated for several years there is the dramatic invasion by several other grasses and herbs. For example, in a forest remnant in Bambalang village, certainly not farmed for a very long time, *Newtonia* dominant trees of 25 to 30m height were observed. It had an understorey of shrubs and herbs such as *Caesalpinia decapetala, Clausena anista*, and *Ouratea fleva.* The creepers included *Clerodendrum umbrellatum, Dalbergia spp*, and *Smilax kraussiana*; and grasses such as *Setaria caudula, Panicum maximum* and *Centotheca lappacea.* This is indicative of the fact that when fallowed for decades of years it can revert to lowland rain forest.

Fig. 2: Model of the shifting cultivation cycle, the process of cropping intensification and post-cultivation vegetation successions

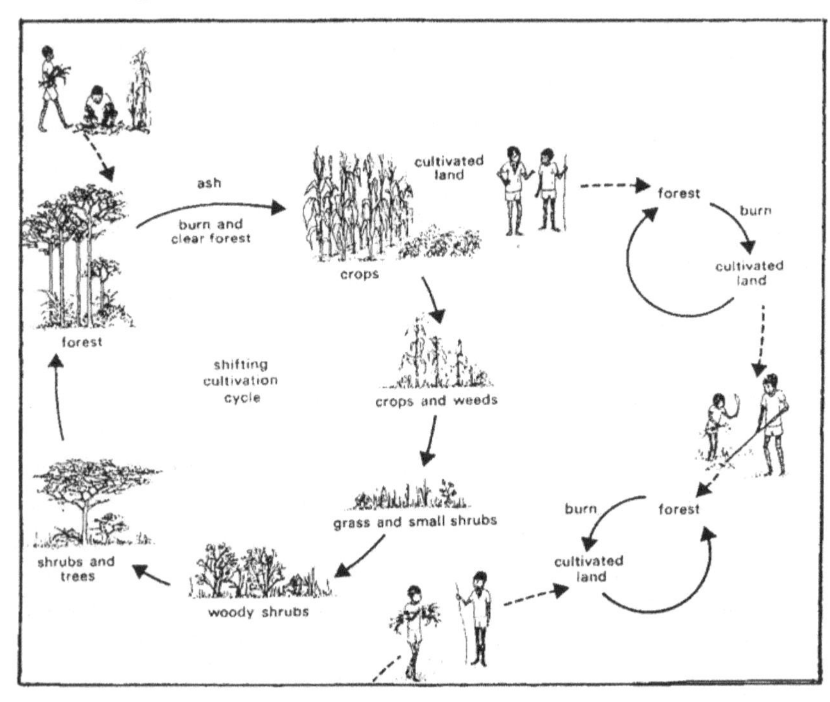

Figure 2 conceptualizes the above successions. Farms clear the climax vegetation by burning the trees and bush. They plant crops in the ash-fertilized soils using the clear-bury in the mounds and burn method. When the fertility declines, or when weed infestation becomes serious, they abandon to move to other locations. After the fields are abandoned, forbs quickly colonize, and gradually give way to woody shrubs and trees as the forest or bush returns. Shifting cultivation can make use of the land for a while but the ecosystem is soon re-established once the human influence ceases. With increasing population pressure on land and the shortening of fallow durations the vegetation is composed of wooded savannas and various post cultivation vegetation successions of grasses and weeds. The typical shrub savannas are *Terminalia* shrub savannas with *Hyparrhenia* grass layer derived from evergreen forest and *Syzigum* guinensis tree savanna which in Bamali has replaced the evergreen forest between Bamali and Bambalang. These shrub savannas are a plagioclimax resulting from the anthropic interruptions of successions towards the climatic climax. Fallows are therefore impoverished by each cultivation cycle as a result of no natural nutrient recycling by a mature forest.

There is the need to enrich fallows by integrating indigenous multipurpose trees in the field. Work in this direction by research has been haphazard (Asah, 1994) with no comprehensive inventory list of socially, ecologically and economically accepted shrubs. Table 4 summarizes the work of Zimmermann (1996), Asah (1994) and Yamoah *et al.*, (1994) in the medium (800m to 1500m above sea level) and high altitude (above 1500m) regions of the Bamenda Highlands.

Table 4: Indigenous woody species feasible for integration in crop production system of the Bamenda Highlands (Asah, 1994; Zimmermann, 1996)

Woody species	*Functions / uses*
Albizia adianthifolia	Fodder, agroforestry, fuel wood, timber
Albizia gummifera	Fodder, agroforestry, fuel wood, timber
Albizia zygia	Fodder, agroforestry, fuel wood, timber
Alchornea cordifolia	Fodder, fuel wood, timber
Anogeissus leiocarpus	Fodder, fuel wood, timber
Bridelia speciosa	Fodder, agroforestry, live fence, medicinal, fuel wood, watershed protection, timber
Caesalpina sp.	Live fence, soil conservation, watershed protection
Calliandra calothynsus	Fodder, agroforestry, live fence, erosion control, fuel wood, N-fixation
Canarium schwei	Food, shells, timber, medicine, gum
Cassia siamea	Fodder, agroforestry, live fence, fuel wood
Cassia spectabilis	Fuel, poles, honey, shade, fodder, watershed protection
Cola acuminata	Edible nuts, soil enrichment, medicinal
Cordia milleny	Poles, timber, carving, medicinal, fodder, watershed protection
Cordia africanus	Fodder, live fence, timber, medicinal, erosion control, fuel wood
Croton macrostachyus	Shade, timber, poles, live fence, watershed protection
Comellina sp.	Live fence, fuelwood, timber, erosion control
Crassocephylum manni	Fodder, agroforestry, live fence, fuel wood

Table 4: Indigenous woody species feasible for integration in crop production system of the Bamenda Highlands (Asah, 1994; Zimmermann, 1996)(Contd)

Woody species	*Functions / uses*
Dracaena sp.	Live fence, traditional uses, erosion control
Daniellia oliveri	Agroforestry, live fence, fuel wood
Datura candida	Live fence, fuel wood
Entanda abyssinica	Live fence, fodder, agroforestry, fuel wood, erosion control, watershed protection
Etandrophragma sp.	Timber, poles, shade, watershed protection
Erythrina sigmoides	N-fixation, live fence, medicinal, fodder
Embelia schimperi	Live fence, fuel wood, medicinal
Ficus glumsa	Live fence, erosion control, fuel wood, watershed protection
Ficus vogellanum	Fodder, live fence, erosion control, fuel wood, watershed protection
Gmelina arborea	Live fence, shade
Grevillea robusta	Shade, honey, reduces soil acidity
Hurungana	Orange dye, poles, medicinal, watershed protection
Khaya grandis	Reforestation, shade, timber, watershed protection
Lassiosiphon glaucus	Paper-making, medicinal, reduces soil acidity, watershed protection
Leucaena leucocephala	Fodder, reduces soil acidity medicinal, fuel wood, watershed protection, live fence, N-Fixation
Leucaena diversifera	Alley planting, green manure, N-fixation, fodder, live fence, fuel wood, watershed protection
Maesa lanceolata	Agroforestry, medicine, erosion control
Maesopsis eminii	Live fence, shade, timber, watershed protection, traditional uses
Markhamia tomentosa	Live fence, erosion control, medicinal, timber, watershed protection
Newtonia buchananii	Reforestation, soil enrichment via foliage
Piliostigma thonningii	Fodder, fuel wood
Podocarpus milanjianus	Reforestation, soil enrichment, timber, watershed protection
Polyscia fulva	Carving, reforestation, soil enrichment, watershed protection
Prenus arifanus	Reforestation, medicinal, soil enrichment, watershed protection
Pseudospodia microcarpa	Agroforestry, fuel wood

Table 4: Indigenous woody species feasible for integration in crop production system of the Bamenda Highlands (Asah, 1994; Zimmermann, 1996)(Contd)

Woody species	*Functions / uses*
Raphia vinifera	Food, handicraft, poles, fuel wood, watershed protection
Schefflera barteri	Live fence, medicinal
Schizolobium	Live fence, erosion control, fuel wood
Sesbania marcrantha	N-Fixation, fodder, erosion control, fuel wood, watershed protection
Spathodea	Reforestation, shade, timber, medicinal, watershed protection
Sorindeia sp.	Reforestation, shade, timber, honey, watershed protection
Terminalia sp.	Live fence, erosion control, fuel wood, shade, timber, watershed, protection
Tephrosia sp.	N-fixation, fuel wood, erosion control, medicinal
Trema orientalis	Reforestation, shade, timber, watershed protection
Vitex diversifolia	Reforestation, carving, timber, medicinal, ink, watershed protection
Vernonia sp.	Food (vegetable), live fence, erosion control

Conclusions

Population pressure in most of Sub-Saharan Africa is such that forest land is being lost to cultivation at a very fast rate; often this leads to the adoption of inappropriate farming systems such as shifting cultivation which does not allow a long enough fallow periods for the land to recover. Apart from the degradation of forest by cultivation, is also the degradation of watersheds, soil and fauna. It is apparent that these are prime areas for the adoption of ecologically integrated land use systems and that agroforestry must from an element of such systems. It has the potential to increase land productivity and to decrease the instability associated with environmental deterioration. The instability is caused by the deployment of ecologically unsound practices in which system outputs exceed inputs. This study recommends the combination of trees with annual crops because of the established micro-ecological benefits such as (Beets, 1989; Reijntjes *et al.*, 1992):

- the reduction of the pressure on forest and therefore more forest trees to protect hill areas from environmental deterioration
- improvement of the water catchment areas;
- more efficient cycling of nutrients by deep rooted trees;

Established on-site ecological benefits include:

- the reduction in the erosive capacity of rainfall, surface runoff, of nutrient leaching and of soil erosion because of tree roots and stems;
- improvement of microclimate; lowering of soil surface temperature and reduction in evaporation of soil moisture through a combination of mulching and shading;
- increase in available soil nutrients through the addition and decomposition of litter fall (nutrient cycling); and
- improvement of soil structure through the constant addition of organic matter from decomposed litter.

Acknowledgements

I thank the late Asah Henry (Agronomist), Kum Sylvester, (Agricultural Engineer) and Bogne Andre (Agricultural Engineer) for field assistance with the survey. The geography students (2005/2006 batch) of E.N.S. Bambili (University of Yaounde I) for assistance in the farm survey.

References

Asah, H. A. (1994) Potentials of multipurpose trees and shrubs in traditional crop-livestock production systems of the Bamenda Highlands of Cameroon. *Proceedings of agroforestry harmonization Workshop*, 4th -7th April, 1994, RCA Bambili, p. 19

Aseh, E. V. (1997) Beyond slashing and burning. The Farmers Voice, No. 019, February 1997, p. 12-15

Bassey, E.E. and Ndenecho, E. N. (2006) A conceptual framework for ecologically and socially sustainable land management and agricultural development in Africa. *Int. Journal of Environment and Sustainable Development, vol. 5, No.3* p. 275 – 286.

Beets, W. C. (1989) The potentials of agroforestry in ACP countries. Tropical Centre for Agricultural and Rural Cooperation, Wageningen, p. 12-15

Chambers, R.; Pacey, A; Thrupp, L. (1989) Farmer First: Farmer innovation and agricultural research, ITP. London.

Champaud, J. (1973) Atlas regional l'Ouest 2. ORSTOM, Yaounde p. 5-16

Harris, F. (1999) Nutrient management strategies of small holder farmers in a short fallow farming system in the north-east Nigeria. *The Geogrphical Journal, vol. 165, No. 3* p. 275-285

Hawkins, P. and Brunt, M. (1965) Soils and ecology of West Cameroon. Report No. 2083, FAO: Rome. p. 492

Kips, P.; Faure, P.; Awah, E.; Kuoh, H.; Sayol, R. and Tchieunteu, F. (1987) Soils, land use and land evaluation of North West Province of Cameroon. FAO/UNDP Report No. 37, IRA/CNS, Ekona.

Lambi, C. (2001) Environmental constraints and indigenous agricultural intensification in Ndop Plain (Upper Nun Valley of Cameroon) In. C. M. Lambi and E.E. Bassey (eds) *Readings in Geography*, Unique Printers, Bamenda, p. 179-190.

Lawton, H. W. and Wike, P. J. (1979) Ancient agricultural systems in dry regions. In A. E. Hall, G. B. Cannel and H.W. Lawton (eds) *Agriculture in semi-arid environments,* New York Springer, p. 1-44.

McHugh, D. (1988) Maize-based farming systems in Ndop Plain of North West Province Cameroon, USAID / IITA/ IRA Bambui. p. 10-26.

Ndenecho, E. N. (2006) Mountain geography and resource conservation. Unique Printers, Bamenda. 184p.

Ndenecho E. N. (2005) Savanization of tropical montane cloud forest in the Bamenda Highlands, Cameroon. *Journal of the Cameroon Academy of Sciences, vol.5*, No. 1 Buea

OTA (1988) Rethinking agriculture in Africa: A role for US Development Assitance. Washington D.C. U.S. Government Printing Agency.

Reijintjes, C. Haverkart, B. and Walters-Bayer, A. (1992) An introduction to low-external input and sustainable agriculture. ILEIA, Leusden. p. 5-25

Sachs, I. (1987) Towards a second green revolution. In B. Glaeser (ed.) The green revolution revisited. Allen and Unwin, London, p. 193 – 198

SEDA (1983) Etude d'identification du sous projet développement rural intégré périmètre Balikumbat-Bambalang, ORSTOM / SEDA, Yaounde. p. 13-16

TAC / CGIAR (1988) Sustainable agricultural production : Implications for international agricultural research. FAO, Rome.

Weiskel, T. E. (1989) The ecological lessons of the past: an anthropology of environmental decline. *The Ecologist, vol, 19, No. 3*. p. 98-103

Yamoah, C.; Ngueguim, C.; Ngong, C.; Cherry, S. (1994) Soil fertility conservation for sustainable crop production: experiences from some highland areas of North – West Cameroon. *Proceeding of agroforestory harmonization workshop*, 4th – 7th April, 1994, RCA Bambili, p. 1-7.

Zimmermann, T. (1996) Watershed resources management in the Western Highlands. *Manual for Technicians*. Helvetas, Bamenda.